AF595040

Non-covalent Interactions in Coordination and Organometallic Chemistry

Non-covalent Interactions in Coordination and Organometallic Chemistry

Editor

Alexander S. Novikov

MDPI • Basel • Beijing • Wuhan • Barcelona • Belgrade • Manchester • Tokyo • Cluj • Tianjin

Editor
Alexander S. Novikov
Institute of Chemistry;
Infochemistry Scientific Center
Saint Petersburg State
University; ITMO University
Saint Petersburg
Russia

Editorial Office
MDPI
St. Alban-Anlage 66
4052 Basel, Switzerland

This is a reprint of articles from the Special Issue published online in the open access journal *Crystals* (ISSN 2073-4352) (available at: www.mdpi.com/journal/crystals/special_issues/organometallic_compounds).

For citation purposes, cite each article independently as indicated on the article page online and as indicated below:

LastName, A.A.; LastName, B.B.; LastName, C.C. Article Title. *Journal Name* **Year**, *Volume Number*, Page Range.

ISBN 978-3-0365-3288-2 (Hbk)
ISBN 978-3-0365-3287-5 (PDF)

Contents

About the Editor vii

Alexander S. Novikov
Non-Covalent Interactions in Coordination and Organometallic Chemistry
Reprinted from: *Crystals* **2020**, *10*, 537, doi:10.3390/cryst10060537 1

Kristina F. Baranova, Aleksei A. Titov, Oleg A. Filippov, Alexander F. Smol'yakov, Alexey A. Averin and Elena S. Shubina
Dinuclear Silver(I) Nitrate Complexes with Bridging Bisphosphinomethanes: Argentophilicity and Luminescence
Reprinted from: *Crystals* **2020**, *10*, 881, doi:10.3390/cryst10100881 5

Younes Hanifehpour, Babak Mirtamizdoust, Sang Woo Joo, Majid Sadeghi-Roodsari and Mehdi Abdolmaleki
A Novel Fishbone-Like Lead(II) Supramolecular Polymer: Synthesis, Characterization, and Application for Producing Nano Metal Oxide
Reprinted from: *Crystals* **2021**, *11*, 335, doi:10.3390/cryst11040335 21

Younes Hanifehpour, Jaber Dadashi and Babak Mirtamizdoust
Ultrasound-Assisted Synthesis and Crystal Structure of Novel 2D Cd (II) Metal–Organic Coordination Polymer with Nitrite End Stop Ligand as a Precursor for Preparation of CdO Nanoparticles
Reprinted from: *Crystals* **2021**, *11*, 197, doi:10.3390/cryst11020197 33

Akbar Ali, Muhammad Khalid, Saba Abid, Muhammad Nawaz Tahir, Javed Iqbal and Muhammad Ashfaq et al.
Green Synthesis, SC-XRD, Non-Covalent Interactive Potential and Electronic Communication *via* DFT Exploration of Pyridine-Based Hydrazone
Reprinted from: *Crystals* **2020**, *10*, 778, doi:10.3390/cryst10090778 45

Alexander S. Novikov
IsoStar Program Suite for Studies of Noncovalent Interactions in Crystals of Chemical Compounds
Reprinted from: *Crystals* **2021**, *11*, 162, doi:10.3390/cryst11020162 65

About the Editor

Alexander S. Novikov

Dr. Alexander S. Novikov is a Senior Researcher at the Institute of Chemistry, Saint Petersburg State University, and an Associate Research Professor and "Computational Chemistry"Group Leader at the Infochemistry Scientific Center, ITMO University, Saint Petersburg, Russia. His research interests include theoretical studies in the following topics: cycloaddition and nucleophilic addition reactions, their mechanisms, driving forces, kinetics and thermodynamics; consideration of the catalysis of hydrocarbons oxidation processes and their conversion to alcohols, ethers, aldehydes, ketones and carboxylic acids; investigations of various unusual types of non-covalentinteractions in organic, organometallic and coordination compounds. He is a co-author of more than 230 articles in reputable international scientific journals.

Google Scholar: <a href=://scholar.google.com/citations?user=n_rBOQcAAAAJ&hl=ru&oi=ao>Alexander S. Novikov - Google</a>

Scopus ID: 50262902200

Web of Science ResearcherID: L-5001-2015

Editorial

Non-Covalent Interactions in Coordination and Organometallic Chemistry

Alexander S. Novikov

Institute of Chemistry, Saint Petersburg State University, Universitetskaya Nab. 7/9, 199034 Saint Petersburg, Russia; ja2-88@mail.ru

Received: 21 June 2020; Accepted: 22 June 2020; Published: 23 June 2020

Abstract: The problem of non-covalent interactions in coordination and organometallic compounds is a hot topic in modern chemistry, material science, crystal engineering and related fields of knowledge. Researchers in various fields of chemistry and other disciplines (physics, crystallography, computer science, etc.) are welcome to submit their works on this topic for our Special Issue "Non-Covalent Interactions in Coordination and Organometallic Chemistry". The aim of this Special Issue is to highlight and overview modern trends and draw the attention of the scientific community to various types of non-covalent interactions in coordination and organometallic compounds. In this editorial, I would like to briefly highlight the main successes of our research group in the field of the fundamental study of non-covalent interactions in coordination and organometallic compounds over the past 5 years.

Keywords: non-covalent interactions; crystal engineering; organometallic compounds; coordination compounds; crystalline materials; supramolecular systems

---

Non-covalent interactions in coordination and organometallic compounds (hydrogen, halogen, chalcogen, pnictogen, tetrel and semi-coordination bonds; agosic and anagosic interactions; stacking, anion-/cation-π interactions; metallophilic interactions, etc.) are topical in modern chemistry, material science, crystal engineering and related fields of knowledge. Both experimental and theoretical methods are widely used for the investigation of the nature and various properties of such weak contacts in gas, liquid and solid states. Non-covalent interactions could be the driving force in the design of smart materials with valuable redox, electronic, mechanical, magnetic and optical properties, and they are promising for the manufacture of LEDs, photovoltaic cells for solar power plants, porous structures, sensors, battery cells and liquid crystals.

In this editorial, I would like to briefly highlight the main successes of our research group in the field of the fundamental study of non-covalent interactions in coordination and organometallic compounds over the past 5 years.

Our group reported the first examples of the unambiguous identification of halogen bonding between metal centers and halocarbons [1] and the application of p-trifluoromethylbenzonitrile moiety for crystal engineering utilizing π-stacking: efficient π-stacking with benzene provides a 2D assembly of *trans*-($PtCl_2$ (*p*-$CF_3C_6H_4CN)_2$) [2]. We published several works about the recognition of the π-hole donor ability of iodofluorobenzenes [3,4]. In particular, we found that structure-directing weak interactions with 1,4-diiodotetrafluorobenzene convert 1D-arrays of ($M^{II}(acac)_2$) species into 3D-networks [5]. The metal-involving halogen bond Ar–I···($d_{z^2}Pt^{II}$) in a platinum acetylacetonate complex was discussed in [6]. We observed that the difference in energy between the two distinct types of chalcogen bonds drives the regioisomerization of binuclear (diaminocarbene)Pd^{II} complexes [7]. The intra-/intermolecular bifurcated chalcogen bonding in the crystal structure of thiazole/thiadiazole derived binuclear (diaminocarbene)Pd^{II} complexes was studied in [8]. The effect of $\mu_{(S,N-H)}Cl$

contacts on the dimerization of Cl(Carbene)Pd^{II} species was discussed in [9]. The effect of π-hole$\cdots\pi$ non-covalent bonding on the conformational stabilization of acyclic diaminocarbene complexes and ligation-enhanced π-hole$\cdots\pi$ interactions involving isocyanides was analyzed in [10]. The (isocyano group π-hole)$\cdots$($d_z{}^2$-M^{II}) interactions at (isocyanide)(M^{II}) complexes, where positively charged metal centers (d^8M = Pt, Pd) act as nucleophiles, were reported in [11]. The (isocyano group)$\cdots$lone pair interactions involving coordinated isocyanides were discussed in [12]. In addition, we reported that intramolecular hydrogen bonding stabilizes *trans*-configuration in mixed carbene/isocyanide Pd^{II} complexes [13]. We fixed the solid state stabilization of unstable hemiketal ligands in copper(II) complexes due to the formation of a intermolecular hydrogen bond network [14] and the stabilization of redox reactive (RNC)Cu^{II} species in crystals via a halogen bond with I_2 [15]. We showed that intramolecular non-covalent B–H$\cdots\pi$(Ph) interaction determines the stabilization of the configuration around the amidrazone C=N bond in *closo*-decaborato amidrazones [16]. We discovered the potential of diiodomethane as a halogen bond donor toward metal-bound halides [17]. Furthermore, other dihalomethanes were also considered as bent bifunctional building blocks for the construction of metal-involving halogen bonded hexagons [18]. In [19], we introduced a concept of four-center nodes: supramolecular synthons based on cyclic halogen bonding. A nice example of halogen contact-induced unusual coloring in the Bi(III) bromide complex due to anion-to-cation charge transfer via Br$\cdots$Br interactions [20] and the electrophilic–nucleophilic dualism of nickel(II) toward Ni$\cdots$I non-covalent interactions, viz. the semicoordination of iodine centers via the electron belt and halogen bonding via the σ-hole [21] were reported. The role of solvent$\cdots$complex halogen bonding in the dramatically enhanced solubility of halide-containing organometallic species in diiodomethane was discussed in [22]. We describe several interesting examples of reverse arene sandwich structures based upon π-hole$\cdots$(M^{II})(d^8M = Pt, Pd) interactions, where positively charged metal centers play the role of a nucleophile [23], and reverse sandwich structures from interplay between lone pair–π-hole atom-directed C$\cdots d_z{}^2$(M) and halogen bond interactions [24]. The halogen bonding-assisted assembly of bromoantimonate(V) and polybromide-bromoantimonate-based frameworks was reported in [25]. The features of halogen bonding in the solid state structures of one- and two-dimensional iodine-rich iodobismuthate(III) complexes were discussed in [26]. The chlorotellurate(IV) supramolecular associates with "trapped" Br_2 via non-covalent halogen$\cdots$halogen interactions were analyzed in [27]. The phenomenon of halogen bonding in isostructural Co(II) complexes with 2-halopyridines was fixed in [28]. The hexaiododiplatinate(II) as a useful supramolecular synthon for halogen bonds involving crystal engineering was considered in [29]. The supramolecular polymers derived from the Pt^{II} and Pd^{II} Schiff base complexes via C(sp^2)–H$\cdots$Hal hydrogen bonding were reported in [30] and [31]. Finally, recently, the supramolecular self-organization via bifurcated $(N–H)_2\cdots$Cl contacts that is responsible for the solid-state fluorescence of 1,2,4-triazole zinc(II) complexes was described in [32].

I hope that other authors will follow my initiative and that readers of this Special Issue of Crystals will have the opportunity to become acquainted with the achievements of researchers in this modern topic.

Funding: This editorial article was written without attracting additional external funding from any scientific foundations.

Conflicts of Interest: The author declares no conflict of interest.

References

1. Ivanov, D.M.; Novikov, A.S.; Ananyev, I.V.; Kirina, Y.V.; Kukushkin, V.Y. Halogen bonding between metal centers and halocarbons. *Chem. Commun.* **2016**, *52*, 5565–5568. [CrossRef] [PubMed]
2. Ivanov, D.M.; Kirina, Y.V.; Novikov, A.S.; Starova, G.L.; Kukushkin, V.Y. Efficient π-stacking with benzene provides 2D assembly of trans-$[PtCl_2(p\text{-}CF_3C_6H_4CN)_2]$. *J. Mol. Struct.* **2016**, *1104*, 19–23. [CrossRef]
3. Novikov, A.S.; Ivanov, D.M.; Bikbaeva, Z.M.; Bokach, N.A.; Kukushkin, V.Y. Noncovalent interactions involving iodofluorobenzenes: The interplay of halogen bonding and weak lp(O)•••π-hole$_{arene}$ interactions. *Cryst. Growth Des.* **2018**, *18*, 7641–7654. [CrossRef]

4. Eliseeva, A.A.; Ivanov, D.M.; Novikov, A.S.; Kukushkin, V.Y. Recognition of π-hole donor ability of iodopentafluorobenzene—A conventional σ-hole donor for crystal engineering involving halogen bonding. *CrystEngComm* **2019**, *21*, 616–628. [CrossRef]
5. Rozhkov, A.V.; Novikov, A.S.; Ivanov, D.M.; Bolotin, D.S.; Bokach, N.A.; Kukushkin, V.Y. Structure-directing weak interactions with 1,4-diiodotetrafluorobenzene convert 1D-arrays of $[M^{II}(acac)_2]$ species into 3D-networks. *Cryst. Growth Des.* **2018**, *18*, 3626–3636. [CrossRef]
6. Rozhkov, A.V.; Ivanov, D.M.; Novikov, A.S.; Ananyev, I.V.; Bokach, N.A.; Kukushkin, V.Y. Metal-involving halogen bond Ar–I•••$[d_z{}^2Pt^{II}]$ in a platinum acetylacetonate complex. *CrystEngComm* **2020**, *22*, 554–563. [CrossRef]
7. Mikherdov, A.S.; Kinzhalov, M.A.; Novikov, A.S.; Boyarskiy, V.P.; Boyarskaya, I.A.; Dar'in, D.V.; Starova, G.L.; Kukushkin, V.Y. Difference in energy between two distinct types of chalcogen bonds drives regioisomerization of binuclear (diaminocarbene)Pd^{II} complexes. *J. Am. Chem. Soc.* **2016**, *138*, 14129–14137. [CrossRef]
8. Mikherdov, A.S.; Novikov, A.S.; Kinzhalov, M.A.; Zolotarev, A.A.; Boyarskiy, V.P. Intra-/intermolecular bifurcated chalcogen bonding in crystal structure of thiazole/thiadiazole derived binuclear (diaminocarbene)Pd^{II} complexes. *Crystals* **2018**, *8*, 112. [CrossRef]
9. Mikherdov, A.S.; Novikov, A.S.; Kinzhalov, M.A.; Boyarskiy, V.P.; Starova, G.L.; Ivanov, A.Y.; Kukushkin, V.Y. Halides held by bifurcated chalcogen–hydrogen bonds. Effect of $\mu_{(S,N-H)}$Cl contacts on dimerization of Cl(Carbene)Pd^{II} species. *Inorg. Chem.* **2018**, *57*, 3420–3433. [CrossRef]
10. Mikherdov, A.S.; Kinzhalov, M.A.; Novikov, A.S.; Boyarskiy, V.P.; Boyarskaya, I.A.; Avdontceva, M.S.; Kukushkin, V.Y. Ligation-enhanced π-hole•••π interactions involving isocyanides. Effect of π-hole•••π non-covalent bonding on conformational stabilization of acyclic diaminocarbene ligands. *Inorg. Chem.* **2018**, *57*, 6722–6733. [CrossRef]
11. Katkova, S.A.; Mikherdov, A.S.; Kinzhalov, M.A.; Novikov, A.S.; Zolotarev, A.A.; Boyarskiy, V.P.; Kukushkin, V.Y. (Isocyano group π-hole)•••$[d_z{}^2\text{-}M^{II}]$ interactions at (isocyanide)$[M^{II}]$ complexes, where positively charged metal centers (d^8M = Pt, Pd) act as nucleophiles. *Chem. Eur. J.* **2019**, *25*, 8590–8598. [CrossRef] [PubMed]
12. Mikherdov, A.S.; Katkova, S.A.; Novikov, A.S.; Efremova, M.M.; Reutskaya, E.Y.; Kinzhalov, M.A. (Isocyano group)•••lone pair interactions involving coordinated isocyanides: Experimental, theoretical and CSD study. *CrystEngComm* **2020**, *22*, 1154–1159. [CrossRef]
13. Mikhaylov, V.N.; Sorokoumov, V.N.; Novikov, A.S.; Melnik, M.V.; Tskhovrebov, A.G.; Balova, I.A. Intramolecular hydrogen bonding stabilizes trans-configuration in a mixed carbene/isocyanide Pd^{II} complexes. *J. Organomet. Chem.* **2020**, *912*, 121174. [CrossRef]
14. Melekhova, A.A.; Novikov, A.S.; Rostovskii, N.V.; Sakharov, P.A.; Panikorovskii, T.L.; Bokach, N.A. Open-chain hemiketal is stabilized by coordination to a copper(II). *Inorg. Chem. Commun.* **2017**, *79*, 82–85. [CrossRef]
15. Bulatova, M.; Melekhova, A.A.; Novikov, A.S.; Ivanov, D.M.; Bokach, N.A. Redox reactive (RNC)Cu^{II} species stabilized in the solid state via halogen bond with I_2. *Z. Kristallogr. Cryst. Mater.* **2018**, *233*, 371–377. [CrossRef]
16. Burianova, V.K.; Bolotin, D.S.; Mikherdov, A.S.; Novikov, A.S.; Mokolokolo, P.P.; Roodt, A.; Boyarskiy, V.P.; Dar'in, D.; Krasavin, M.; Suslonov, V.V.; et al. Mechanism of generation of closo-decaborato amidrazones. Intramolecular non-covalent B–H•••π(Ph) interaction determines stabilization of the configuration around the amidrazone C=N bond. *New J. Chem.* **2018**, *42*, 8693–8703. [CrossRef]
17. Novikov, A.S.; Ivanov, D.M.; Avdontceva, M.S.; Kukushkin, V.Y. Diiodomethane as a halogen bond donor toward metal-bound halides. *CrystEngComm* **2017**, *19*, 2517–2525. [CrossRef]
18. Kashina, M.V.; Kinzhalov, M.A.; Smirnov, A.S.; Ivanov, D.M.; Novikov, A.S.; Kukushkin, V.Y. Dihalomethanes as bent bifunctional XB/XB-donating building blocks for construction of metal-involving halogen bonded hexagons. *Chem. Asian J.* **2019**, *14*, 3915–3920. [CrossRef]
19. Kryukova, M.A.; Ivanov, D.M.; Kinzhalov, M.A.; Novikov, A.S.; Smirnov, A.S.; Bokach, N.A.; Kukushkin, V.Y. Four-center nodes: Supramolecular synthons based on cyclic halogen bonding. *Chem. Eur. J.* **2019**, *25*, 13671–13675. [CrossRef]
20. Adonin, S.A.; Gorokh, I.D.; Novikov, A.S.; Abramov, P.A.; Sokolov, M.N.; Fedin, V.P. Halogen contacts-induced unusual coloring in Bi(III) bromide complex: Anion-to-cation charge transfer via Br•••Br interactions. *Chem. Eur. J.* **2017**, *23*, 15612–15616. [CrossRef]

21. Bikbaeva, Z.M.; Ivanov, D.M.; Novikov, A.S.; Ananyev, I.V.; Bokach, N.A.; Kukushkin, V.Y. Electrophilic–nucleophilic dualism of nickel(II) toward Ni•••I non-covalent interactions: Semicoordination of iodine centers via electron belt and halogen bonding via σ-Hole. *Inorg. Chem.* **2017**, *56*, 13562–13578. [CrossRef] [PubMed]
22. Kinzhalov, M.A.; Kashina, M.V.; Mikherdov, A.S.; Mozheeva, E.A.; Novikov, A.S.; Smirnov, A.S.; Ivanov, D.M.; Kryukova, M.A.; Ivanov, A.Y.; Smirnov, S.N.; et al. Dramatically enhanced solubility of halide-containing organometallic species in diiodomethane: The role of solvent•••complex halogen bonding. *Angew. Chem. Int. Ed.* **2018**, *57*, 12785–12789. [CrossRef] [PubMed]
23. Rozhkov, A.V.; Krykova, M.A.; Ivanov, D.M.; Novikov, A.S.; Sinelshchikova, A.A.; Volostnykh, M.V.; Konovalov, M.A.; Grigoriev, M.S.; Gorbunova, Y.G.; Kukushkin, V.Y. Reverse arene sandwich structures based upon π-hole•••[M^{II}](d^8M = Pt, Pd) interactions, where positively charged metal centers play the role of a nucleophile. *Angew. Chem. Int. Ed.* **2019**, *58*, 4164–4168. [CrossRef] [PubMed]
24. Baykov, S.V.; Filimonov, S.I.; Rozhkov, A.V.; Novikov, A.S.; Ananyev, I.V.; Ivanov, D.M.; Kukushkin, V.Y. Reverse sandwich structures from interplay between lone pair–π-hole atom-directed C•••$d_z{}^2$[M] and halogen bond interactions. *Cryst. Growth Des.* **2020**, *20*, 995–1008. [CrossRef]
25. Adonin, S.A.; Bondarenko, M.A.; Novikov, A.S.; Abramov, P.A.; Plyusnin, P.E.; Sokolov, M.N.; Fedin, V.P. Halogen bonding-assisted assembly of bromoantimonate(V) and polybromide-bromoantimonate-based frameworks. *CrystEngComm* **2019**, *21*, 850–856. [CrossRef]
26. Adonin, S.A.; Usoltsev, A.N.; Novikov, A.S.; Kolesov, B.A.; Fedin, V.P.; Sokolov, M.N. One- and two-dimensional iodine-rich iodobismuthate(III) complexes: Structure, optical properties and features of halogen bonding in the solid state. *Inorg. Chem.* **2020**, *59*, 3290–3296. [CrossRef]
27. Usoltsev, A.N.; Adonin, S.A.; Novikov, A.S.; Abramov, P.A.; Sokolov, M.N.; Fedin, V.P. Chlorotellurate(IV) supramolecular associates with trapped. Br_2: Features of non-covalent halogen•••halogen interactions in crystalline phases. *CrystEngComm* **2020**, *22*, 1985–1990. [CrossRef]
28. Adonin, S.A.; Bondarenko, M.A.; Novikov, A.S.; Sokolov, M.N. Halogen bonding in isostructural Co(II) complexes with 2-halopyridines. *Crystals* **2020**, *10*, 289. [CrossRef]
29. Eliseeva, A.A.; Ivanov, D.M.; Novikov, A.S.; Rozhkov, A.V.; Kornyakov, I.V.; Dubovtsev, A.Y.; Kukushkin, V.Y. Hexaiododiplatinate(II) as a useful supramolecular synthon for halogen bond involving crystal engineering. *Dalton Trans.* **2020**, *49*, 356–367. [CrossRef]
30. Tskhovrebov, A.G.; Novikov, A.S.; Odintsova, O.V.; Mikhaylov, V.N.; Sorokoumov, V.N.; Serebryanskaya, T.V.; Starova, G.L. Supramolecular polymers derived from the Pt^{II} and Pd^{II} Schiff base complexes via C(sp^2)–H•••Hal hydrogen bonding: Combined experimental and theoretical study. *J. Organomet. Chem.* **2019**, *886*, 71–75. [CrossRef]
31. Repina, O.V.; Novikov, A.S.; Khoroshilova, O.V.; Kritchenkov, A.S.; Vasin, A.A.; Tskhovrebov, A.G. Lasagna-like supramolecular polymers derived from the Pd^{II} osazone complexes via C(sp^2)–H•••Hal hydrogen bonding. *Inorg. Chim. Acta* **2020**, *502*, 119378. [CrossRef]
32. Yunusova, S.N.; Novikov, A.S.; Khoroshilov, O.V.; Kolesnikov, I.E.; Demakova, M.Y.; Bolotin, D.S. Solid-state fluorescent 1,2,4-triazole zinc(II) complexes: Self-organization via bifurcated $(N–H)_2$•••Cl contacts. *Inorg. Chim. Acta* **2020**, *510*, 119660. [CrossRef]

Article

Dinuclear Silver(I) Nitrate Complexes with Bridging Bisphosphinomethanes: Argentophilicity and Luminescence

Kristina F. Baranova [1,2], Aleksei A. Titov [1,*], Oleg A. Filippov [1,3], Alexander F. Smol'yakov [1], Alexey A. Averin [4] and Elena S. Shubina [1,*]

[1] A. N. Nesmeyanov Institute of Organoelement Compounds, Russian Academy of Sciences, Vavilov Str., 28, 119991 Moscow, Russia; krisbar99@gmail.com (K.F.B.); h-bond@ineos.ac.ru (O.A.F.); rengenhik@gmail.com (A.F.S.)

[2] Faculty of Chemistry, Lomonosov Moscow State University, 1-3 Leninskie Gory, 119991 Moscow, Russia

[3] Shemyakin-Ovchinnikov Institute of Bioorganic Chemistry, Russian Academy of Sciences, Miklukho-Maklaya St., 16/10, 117997 Moscow, Russia

[4] A. N. Frumkin Institute of Physical Chemistry and Electrochemistry, Russian Academy of Sciences, Leninsky prosp. 31/4, 199071 Moscow, Russia; alx.av@yandex.ru

* Correspondence: tit@ineos.ac.ru (A.A.T.); shu@ineos.ac.ru (E.S.S.); Tel.: +7-499-135-1871 (E.S.S.)

Received: 3 July 2020; Accepted: 25 September 2020; Published: 29 September 2020

Abstract: Two silver nitrate complexes with bisphosphines were obtained and characterized: $[Ag(dcypm)]_2(NO_3)_2$ (**1**; dcypm = bis(dicyclohexylphosphino)methane) and $[Ag(dppm)]_2(Me_2Pz^H)_n(NO_3)_2$ (n = 1, **2a**; n = 2, **2b**; dppm = bis(diphenylphosphino)methane, Me_2Pz^H = 3,5-dimethylpyrazole). The steric repulsions of bulky cyclohexyl substituents prevent additional ligand coordination to the silver atoms in **1**. Compounds obtained feature the bimetallic eight-member cyclic core $[AgPCP]_2$. The intramolecular argenthophilic interaction (d(Ag···Ag) = 2.981 Å) was observed in complex **1**. In contrast, the coordination of pyrazole led to the elongation of Ag···Ag distance to 3.218(1) Å in **2a** and 3.520 Å in **2b**. Complexes 1 and 2a possess phosphorescence both in the solution and solid state. Time-dependent density-functional theory (TD-DFT) calculations demonstrate the origin of their different emission profile. In the case of **1**, upon excitation, the electron leaves the Ag–P bonding orbital and locates on the intramolecular Ag···Ag bond (metal-centered character). Complex **2a** at room temperature exhibits a phosphorescence originating from the $^3(M + L^{P+N})L^{Ph}CT$ state. At 77 K, the photoluminescence spectrum of complex **2a** shows two bands of two different characters: $^3(M + L^{P+N})L^{Ph}CT$ and $^3LC^{Ph}$ transitions. The contribution of Ag atoms to the excited state in both complexes **2a** and **2b** decreased relative to **1** in agreement with the structural changes caused by pyrazole coordination.

Keywords: silver phosphine; pyrazole; luminescence; TD-DFT

1. Introduction

The non-covalent metal–metal interactions play a significant role in the emission [1,2] or catalytic [3–5] properties of the d^{10} metal complexes. Dinuclear or trinuclear coinage metal complexes with bridging bis- or trisphosphine ligands represent a unique class of molecules occupying a position between mononuclear complexes and nanoparticles. They have remarkable chemical and physicochemical properties due to the presence of metal–metal bonds of various energy [6]. For example, di- and tetranuclear silver nitrates with bis(diphenylphosphino)methane are known since 1983 [7]. The metal-centered (MC) nature of emission is well known for coinage metal complexes with bridging phosphines. Already in 1989, Che et al. showed on an example of gold dimer $[Au_2(dppm)_2]^{2+}$

that shortened Au–Au distances participate in electron transitions [8]. Then, on the example of silver dinuclear complexes with bisphosphines $[Ag(dcypm)]_2X_2$ (X = PF_6^-, and $CF_3SO_3^-$, or CN^-; dcypm = bis(dicyclohexylphosphino)methane) [9,10] and $[Ag(dmb)]_2X_2$ complexes (X = Cl, Br, I; dmb = 1,8-diisocyano-ρ-menthane) [11], it has been shown that UV–Vis (Ultraviolet–visible) absorption band of complexes to d–p transition from intramolecular argentophilic interaction. The metallophilic nature of this band corresponding to the stretching Ag–Ag vibration (80 cm^{-1}) was proven by Raman spectroscopy with excitation at the UV maximum [9]. The same behavior has also been observed for metallothioneins [11,12] and nanoclusters [13,14]. The interaction between two coinage metal ions has been investigated by spectral as well as computational methods [15,16]. Recently, using the example of Au/Ag di- and trinuclear complexes with diphosphanyl NHC ligands, this phenomenon was explained experimentally and theoretically [17].

Surprisingly, despite the wide structural diversity of coinage metals and bisphosphine complexes, there is practically no information about silver(I) nitrate complexes. Herein, we present the synthesis, structures, photophysical properties of dinuclear silver(I) bisphosphines $[Ag(dcypm)]_2(NO_3)_2$ (**1**) (dcypm = bis(dicyclohexylphosphino)methane) and $[Ag(dppm)]_2(Me_2Pz^H)_n(NO_3)_2$ (n = 1, **2a**; n = 2, **2b**) (dppm = diphenylphosphino)methane) with 3,5-dimethylpyrazole (Scheme 1). The influence of Ag···Ag distances or the ligand at silver atoms on the emission properties will be discussed involving the theoretical interpretation.

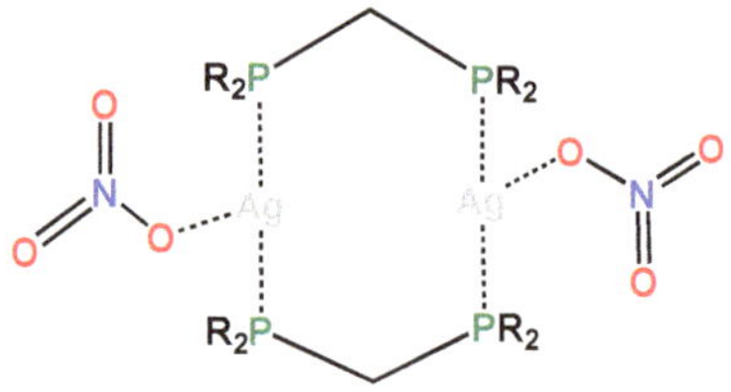

R=Ph or Cy

Scheme 1. Representation of central $[MPCP]_2$ core for the complexes studied (R = Cy (**1**), Ph (**2**)).

2. Materials and Methods

^1H and $^{31}P\{^1H\}$ nuclear magnetic resonance (NMR) measurements were carried out on a Bruker Avance 400 spectrometer. Infrared (IR) spectra were collected on a Shimadzu IR Prestige-21 FT-IR (Fourier-transform infrared spectroscopy) spectrometer using KBr pellets. The photoluminescence spectra and lifetime measurements of the phosphors were recorded at 77 K and 298 K on a Fluorolog-3 spectrofluorometer system (HORIBA Jobin-Yvon, the excitation source was a 450 W xenon lamp with Czerny–Turner double monochromators, and the registration channels were a R928 photomultiplier, while a 150 W pulsed xenon lamp was used for lifetime measurements). The powders for these measurements were packed in quartz capillaries. The phosphorescence quenching curves were analyzed using the FluoroEssence™ software for calculation of the phosphorescence lifetime values. The UV–Vis spectra were measured on a Cary50 spectrometer.

X-ray diffraction study. Single-crystal X-ray diffraction experiments for complexes **1**, **2a**, and **2a/2b** were carried out with a Bruker SMART APEX II diffractometer. The structures were solved by the direct method and refined in anisotropic approximation for non-hydrogen atoms. Hydrogen atoms of methyl, methylene, and aromatic fragments were calculated according to those idealized geometries and refined with constraints applied to C–H and N–H bond lengths and equivalent displacement parameters ($U_{eq}(H) = 1.2U_{eq}(X)$, where X is the central atom of the XH_2 group; and $U_{eq}(H) = 1.5U_{eq}(Y)$, Y—central atom of YH_3 group). All structures were solved with the ShelXT program [18] and refined with the ShelXL program [19]. Molecular graphics were drawn using the OLEX2 program [20]. The cyclohexyl fragment in complex **1**, Me_2Pz^H ligand in complex **2a**, and in the second molecule in co-crystal **2a/2b** were disordered, and the occupancy of positions of disordered levels was 0.5. Nitrate anions in the second molecule (**2a**) in co-crystal **2a/2b** were disordered, and the occupancy of positions of disordered levels were 0.3 and 0.7. CCDC 2013947, 2013948, and 2033764 contained the supplementary crystallographic data for complexes **1**, **2a**, and **2a/2b**. These data can be obtained free of charge from the Cambridge Crystallographic Data Center via https://www.ccdc.cam.ac.uk/structures. Crystal data and structure refinement parameters are presented in Table 1.

Table 1. Crystallographic data for complexes **1**, **2a**, and **2a/2b**.

	1	**2a**	**2a/2b**
Empirical Formula	$C_{25}H_{46}AgNO_3P_2$	$C_{55}H_{53}Ag_2N_4O_6P_4$	$C_{117}H_{116}Ag_4Cl_2N_5O_6P_4$
Formula weight	578.44	1205.63	2675.24
Diffractometer	Bruker SMART APEX CCD	Bruker SMART APEX CCD	Bruker SMART APEX CCD
Scan mode	ϕ and ω scans	ϕ and ω scans	ω and ϕ scans
Anode [Wavelength, Å]	MoKα [0.71073] sealed tube	MoKα [0.71073] sealed tube	MoKα [0.71073] sealed tube
Crystal Dimensions, mm	0.1 × 0.15 × 0.36	0.08 × 0.11 × 0.15	0.08 × 0.12 × 0.44
Crystal color	colorless	colorless	colorless
Crystal system	tetragonal	monoclinic	triclinic
a, Å	29.8637(11)	10.940(2)	10.9934(13)
b, Å	29.8637(11)	19.805(4)	11.9540(14)
c, Å	12.6441(5)	11.719(2)	22.858(3)
α, deg	90	90	77.727(2)
β, deg	90	91.087(4)	82.796(2)
γ, deg	90	90	85.688(2)
Volume, $Å^3$	11276.5(9)	2538.5(9)	2908.5(6)
Density, g cm^{-3}	1.363	1.577	1.527
Temperature, K	120	120	120
Tmin/Tmax	0.7056/0.7461	0.6339/0.7460	0.6141/0.7459
μ, mm^{-1}	0.853	0.953	0.930
Space group	$I4_1/a$	$P2_1/n$	P1
Z	16	2	2
F(000)	4864	1226	1360
Reflections collected	54587	23611	37353
Independent reflections	5546	4995	16342
Reflections (I > 2σ (I))	4576	3545	9421
Parameters	234	308	740
R_{int}	0.0350	0.1053	0.0860
2θmin–2θmax	2.728–51.998	4.040–52.000	3.670–59.410
wR_2 (all reflections)	0.2032	0.1571	0.1221
R_1 (I > σ (I))	0.0717	0.0866	0.0561
GOF	1.027	1.220	0.972
ρ_{min}/ρ_{max}, $eÅ^{-3}$	−1.042/1.733	−1.338/0.927	−1.011/1.385

Computational details. Full geometry optimizations of complexes **1** and **2** were performed using the PBE0 functional [21] and SVP (split valence polarization) basis set [22,23] without any symmetry restrictions using Gaussian09 software [24]. Optimized geometries of **1**, **2a**, and **2b** reasonably reproduced the X-ray geometry (Figure S14 in the Supplementary Materials). The equilibrium geometry for **1** and **2b** had a C_i symmetry point group (as it was found by X-ray analysis), leading to the symmetry forbidden electronic transitions, so only transitions of A_u symmetry were analyzed. For complex **2a**, the second pyrazole molecule was left, but rotated to be in the Me_2PzH ligand position of a neighbor complex in the crystal. Herewith, that molecule remained non-bonded by the Ag–N bond, but its presence helps to keep the complex similar to the X-ray characterized one. Vertical singlet and triplet excitations were calculated by TD-DFT on the same level of theory. Analysis of the excited states was done with the Multiwfn program [25]. The similarity of the excited states was analyzed by the approach suggested by Chen [26].

$$s_{H/E} = 1 - \frac{\sum_i |a_i - b_i|}{2}$$

where a_i and b_i are contributions of the i atom to the hole (s_H) or electron (s_E) in the states a and b.

Synthesis. All operations were performed with Schlenk techniques under a dry argon atmosphere. Commercially available solvents and ligands were used without additional purification.

[Ag(dcypm)]$_2$(NO$_3$)$_2$ (1). Mixture of $AgNO_3$ (50 mg, 0.294 mmol) and bis(dicyclohexylphosphino) methane (120 mg, 0.294 mmol) was stirred in 5 mL of CH_2Cl_2 at room temperature for 3 h. A white solid was afforded upon precipitation by n-hexane. The precipitate formed was filtered, washed with 10 mL of acetone, and dried in vacuum. Yield 91%. NMR (CD_2Cl_2, ppm) ^{1}H: δ = 1.19–2.13 ppm (m, 88 H^{Cy} and 4 CH_2), $^{31}P\{^1H\}$: δ = 28.8 (m, 4P). Anal. calcd/found for $C_{50}H_{92}Ag_2N_2O_6P_4$: C, 51.91/52.01; H, 8.02/8.08; N, 2.42/2.40.

[Ag(dppm)]$_2$(Me$_2$Pz^H)(NO$_3$)$_2$ (2a). The suspension of 50 mg of $AgNO_3$ (0.294 mmol) and 17.3 mg of 3,5-dimethylpyrazole (0.180 mmol) was stirred in 10 mL of acetone for the one hour until a slurry precipitate was formed, and then the solution of dppm (113 mg, 0.294 mmol) in 3 mL of acetone was added. The reaction mixture was stirred overnight at room temperature. The precipitate formed was filtered off and washed with 10 mL of CH_2Cl_2. The solution obtained was added to 20 mL of hexane, and the mixture was kept in the refrigerator at ca. 5 °C for one hour. The precipitate formed was filtered off, washed with hexane, and both solids were combined and dried under reduced pressure. Complex **2a** was obtained by slow crystallization at room temperature from boiling CH_3CN solution of small portions of these solids. Yield 68%. NMR (CD_3CN, ppm) ^{1}H: δ = 2.07 ppm (s, 6H, Me^{Pz}), 3.71 ppm (m, 4H, CH_2^{dppm}), 5.74 ppm (s, 1H, CH^{Pz}), 7.20–7.44 (m, 40H, Ph^{dppm}), $^{31}P\{^1H\}$: δ = 7.5 (m, 4P); IR (KBr, cm^{-1}): ν 3201 (νNH), 3138, 3050, 2922, 2876 (νCH), 1580. Anal. calcd/found for $C_{55}H_{54}Ag_2N_4O_6P_4$: C, 54.75/54.07; H, 4.51/4.24; N, 4.64/4.60.

3. Results and Discussion

3.1. Synthetic Procedures

Complex 1 was obtained by the interaction of $AgNO_3$ with bis(dicyclohexylphosphino)methane (dcypm) in dichloromethane (DCM) and subsequent precipitation with hexane. Moreover, the addition of dcypm to the solution of the complex [Ag(Me$_2$Pz^H)]NO$_3$ generated in situ only led to complex **1** with high yield. This suggests that the steric effect of bulk cyclohexyl substituents makes impossible the pyrazole coordination to the silver atoms. In contrast, mixing $AgNO_3$ with 3,5-dimethylpyrazole and less sterically demanding bis(diphenylphosphino)methane (dppm) in acetone led to the formation of the precipitate. The elemental analysis and ^{1}H NMR spectrum demonstrated that the composition of the bulk solids corresponded to the complex [Ag$_2$(Me$_2$PzH)(dppm)]$_2$(NO$_3$)$_2$ (**2b**) containing two pyrazole molecules per [Ag(dppm)]$_2$(NO$_3$)$_2$ moiety (Figures S19 and S20). Unfortunately, reprecipitation or crystallization of this complex at low temperatures led to the partial loss of pyrazole molecules, resulting in a mixture of complexes [Ag(dppm)]$_2$(Me$_2$Pz^H)(NO$_3$)$_2$ (**2a**) [Ag$_2$(Me$_2$Pz^H)(dppm)]$_2$(NO$_3$)$_2$ (**2b**). In a pure form, only complex **2a** could be obtained by slow crystallization from boiling acetonitrile solution at room temperature. It should be noted that the elimination of the second 3,5-dimethylpyrazole molecule was not observed.

3.2. Crystal Structure of Complexes

The structures of compounds obtained were proved by single-crystal X-ray diffraction. The crystallographic data and structure refinement details are summarized in Table 1, and the relevant bond distances and angles are collected in Table 2.

Table 2. Selected geometric parameters (bond lengths in Å, angles in degrees) for complexes **1** and **2a**.

Bonds Lengths				Angles			
1		2		1		2	
Ag1-P1	2.396(2)	Ag1-P1	2.403(2)	Ag1-Ag1-P2	82.90(4)	Ag1-Ag1-P2	81.59(5)
Ag1-O1	2.557(9)	Ag1-O1	2.554(6)	Ag1-P1-C1	111.1(2)	Ag1-O1-N3	129.8(5)
Ag1-Ag1	2.9810(7)	Ag1-N1	2.443(6)	Ag1-P1-C2	115.5(2)	Ag1-N1-N2	124.7(5)
Ag1-P2	2.428(2)	Ag1-Ag1	3.218(1)	Ag1-P1-C8	113.2(3)	P1-Ag1-O1	126.5(1)
P1-C1	1.838(8)	Ag1-P2	2.442(2)	Ag1-O1-N1	106.9(5)	P1-Ag1-N1	98.6(2)
P1-C2	1.850(6)	P1-C4	1.846(8)	P1-Ag1-O1	119.1(2)	P1-Ag1-Ag1	88.90(5)
P1-C8	1.85(1)	P1-C17	1.801(8)	P1-Ag1-Ag1	94.21(4)	P1-Ag1-P2	151.61(8)
P2-C1	1.827(7)	P1-C23	1.828(8)	P1-C1-P2	114.4(3)	O1-Ag1-N1	84.1(2)
P2-C17	1.84(1)	P2-C4	1.822(8)	P1-Ag1-P2	152.22(6)	O1-Ag1-Ag1	91.5(1)
P2-C23	1.844(6)	P2-C5	1.827(8)	O1-Ag1-Ag1	74.4(2)	O1-Ag1-P2	80.7(1)
O1-N1	1.27(1)	P2-C11	1.804(8)	O1-Ag1-P2	86.9(2)	N1-Ag1-Ag1	172.5(2)
O2-N1	1.244(8)	P2-Ag1	2.442(2)			N1-Ag1-P2	91.7(2)
O3-N1	1.12 (1)	O1-N3	1.25(1)			N2-N1-C1	101.8(6)
		O2-N3	1.24(1)				
		O3-N3	1.24(1)				
		N1-N2	1.31(1)				
		N1-C1	1.50(1)				
		N1-C2	1.58(1)				
		N2-C3	1.34(1)				
		N2-C2	1.02(1)				

Colorless crystals of complex **1** was obtained by slow evaporation of their DCM/hexane (*v*/*v* = 2:1) solutions at ca. 5 °C. Complex **1** features a chair form of the eight-member cyclic $[AgPCP]_2$ core (Figure 1), in which silver atom is coordinated to four atoms (two P^{dppm} (2.396(2) and 2.429(2) Å), $O^{Nitrate}$ (2.558(9) Å) and Ag (2.9812(7) Å), forming the distorted triangular pyramidal or tetrahedron environment (Figure 2). The angles ∠P1-Ag1-P2, ∠P1-Ag1-O1, ∠P1-Ag1-Ag1, ∠O1-Ag1-Ag1 were 152.22(6)°, 119.1(2)°, 94.21(4)°, and 74.4(2)°, respectively. There were only several examples of the similar $[Ag(dcypm)]_2^{2+}$ core with PF_6^-, $CF_3SO_3^-$ [9], and CN^- [10] counter ions. Observed Ag–Ag bond length (2.9812(7) Å) was in the range of shared argentophilic interactions [27–29] being non-significantly longer than that present in the literature (av. 2.93 Å). The central core $[MPCP]_2$ was similar for dinuclear silver bisphosphines, but additional κ^1 coordination of the nitrate anion led to a significant distortion of the central cycle. Silver atoms as well as CH_2 fragments lie out the P_4 plane for 0.557 and 0.607 Å, respectively. For example, the same parameters for complex with CN^- were 0.415 and 0.575 Å [10]. These structural peculiarities reflect the influence of the nitrate anion on the electronic properties of complexes (*vide infra*). The Ag–P bond lengths (2.396(2) and 2.429(2) Å) were typical for the silver bis-phosphines.

Supramolecular packing of **1** was realized via the network of weak C^{Cy}-H···$O^{Nitrate}$ interactions of neighboring molecules (2.316 Å, ∠O-H-C = 156.1°; Figure 2) [30,31].

In contrast, in the case of interaction of the silver nitrate with dppm ligand and 3,5-dimethylpyrazole, complexes of different compositions were observed. As demonstrated above, the interaction led to the formation of the precipitate. Analysis of this solid indicates that it mainly contained one Ag atom per one dppm ligand and one Me_2Pz^H molecule. However, varying the crystallization conditions allowed for two different types of complexes. The solubility of the solids obtained was very low, and crystallization was performed from significantly diluted solutions (mM concentrations). The most stable one was complex **2a** containing one pyrazole molecule, crystals that could be obtained at different temperatures (−10, 0, or 25 °C) and from different solvents (MeOH, CH_3CN, and DCM/hexane mixture). Slow crystallization from the boiling CH_3CN at room temperature was the best condition for obtaining pure complex **2a**. Complex **2b** was obtained only as a co-crystal with **2a**, representing two independent molecules with one and two Me_2Pz^H molecules. This sample was crystallized in an NMR tube after a ^1H NMR experiment in CD_2Cl_2, which eventually showed the 1/1 ratio of

Me_2Pz^H/dppm in solution. These data demonstrate that crystallization from a diluted solution led to the elimination of one pyrazole from the complex $[Ag(dppm)Me_2Pz^H]_2(NO_3)_2$ (**2b**), resulting in a more stable $[Ag(dppm)]_2(Me_2Pz^H)(NO_3)_2$ (**2a**) complex.

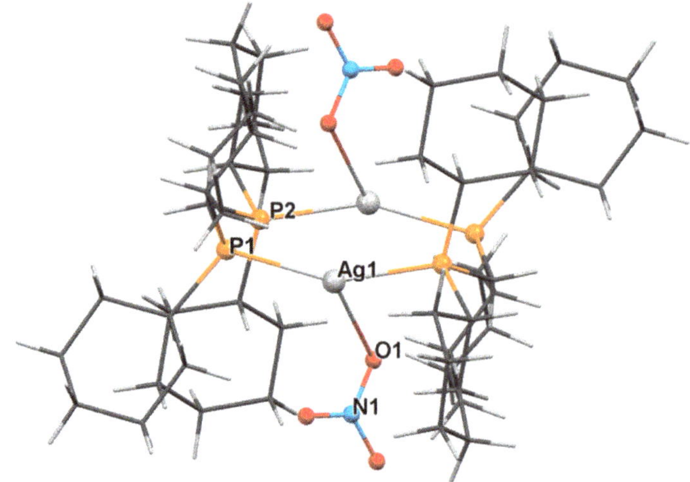

Figure 1. X-ray diffraction (XRD) structure of complex **1** (carbon and hydrogen atoms are shown as sticks).

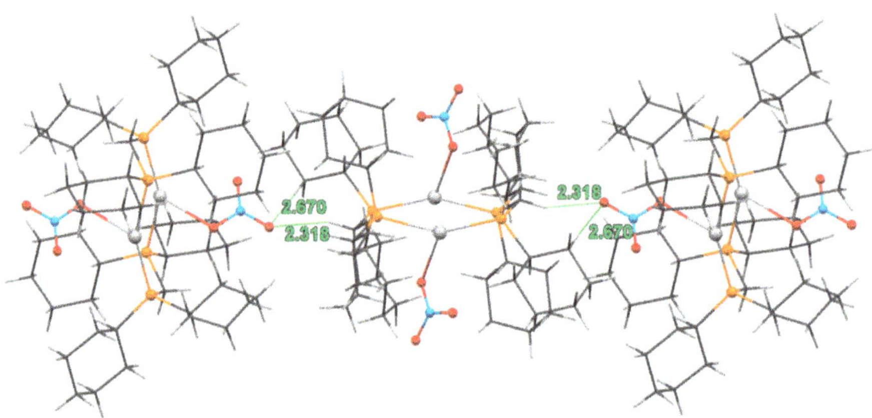

Figure 2. Supramolecular packing of **1**.

Complex **2a** crystallizes in a chair form of the central $[AgPCP]_2$ cycle (Figure 3). There was only one molecule of pyrazole coordinated with one of the silver atoms in the crystal. The complex in the crystal was in the superposition of two possible isomers in which the pyrazole ligand coordinates by nitrogen to both silver atoms (occupation 0.5). This behavior leads to the presence of a pseudo inversion center at the center of the Ag–Ag distance. Interaction with Me_2Pz^H ligand led to the pentacoordinate environment distorted trigonal bipyramidal of one silver atom (two P^{dppm} (2.403(2) and 2.442(2) Å), N^{Pz} (2.446(6) Å), $O^{Nitrate}$ (2.554(6) Å), and Ag (3.218(1) Å) (Figure 3). The angles P1-Ag1-P2, P1-Ag1-O1, and P1-Ag1-N1 were 151.61(8)°, 126.5(1)°, and 98.6(2)°, respectively. The presence of the 3,5-dimethylpyrazole ligand led to the κ^1 coordination of the nitrate to silver atoms in contrast to the known $[Ag(dppm)]_2(NO_3)_2$ complex [7,32], which possesses the κ^2 coordination mode of the nitrate anion. Moreover, the coordination of pyrazole led to the elongation of the Ag–Ag distance to 3.218 Å, which was significantly longer than that for $[Ag(dppm)]_2(NO_3)_2$ (3.090–3.110 Å) [7,32]. The Ag–Ag distance was in the range of closed-shell metallophilic interactions [27]. Silver atoms as well as CH_2 fragments, were located on both sides of the P_4 plane lying out at 0.594 and 0.720 Å, respectively, which determines the chair-configuration of the central M_2P_4 core. It should be noted that Tiekink [32] also reported the same configuration for $[Ag(dppm)(NO_3)]_2$ with a more distorted central core in contrast to the boat form obtained by Bau [7]. The Ag–P bonds lengths in **2** (2.403(2) and 2.442(2) Å) are typical for the silver bis-phosphines complexes. The Ag–N^{Pz} bond length (2.446(4) Å) is typical for the silver phosphine complexes with donor aromatic nitrogen ligands [33,34]. The complex was stabilized via $O_1^{Nitrate}$...H–N^{Pz} intramolecular hydrogen bond (1.702 Å, ∠O-H-N = 165.9°) [30,31].

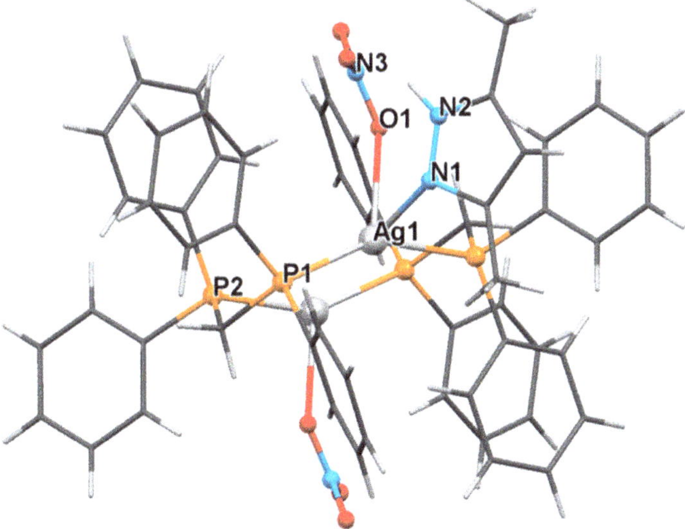

Figure 3. XRD structure of the complex **2** (carbon and hydrogen atoms are shown as sticks).

Complex **2b** was obtained only in the form of a co-crystal of two molecules of general motif $[Ag(dppm)]_2(NO_3)_2$ with one and two pyrazoles, respectively (Figure 4). The part that contained one molecule of pyrazole was generally identical to compound **2a** with non-significant differences in bond lengths and angles. The second part was a dinuclear complex **2b** with two molecules of 3,5-dimethylpyrazole coordinated to both silver atoms. Complex **2b** also crystallized in a chair form of the central $[AgPCP]_2$ cycle. Moreover, there was the inversion center of this fragment in the middle of the Ag–Ag distance. The second pyrazole molecule's coordination led to the significant elongation of the Ag–Ag distance (3.520 Å). As a result, the presence of one or two coordinated ligands to the silver atoms in $[Ag(dppm)]_2$ cores led to the elongation of Ag–Ag distances and the absence of the shared intramolecular argenthophilic interactions. The Ag–P (2.446(1) and 2.422(1) Å) bond lengths were similar to those observed in complexes **1** and **2a**. The interaction of silver atoms with pyrazoles in **2b** was non-significantly stronger than that for **2a** in accordance with the correlation of the shortened Ag–N^{Pz} bond lengths (2.446(1) Å for **2a** and 2.381(4) Å for **2b**). In contrast, the coordination of addition pyrazole led to the significant elongation of the Ag–$O^{Nitrate}$ bonds (2.554(6) Å for **2a** and 2.674 (3) Å for **2b**).

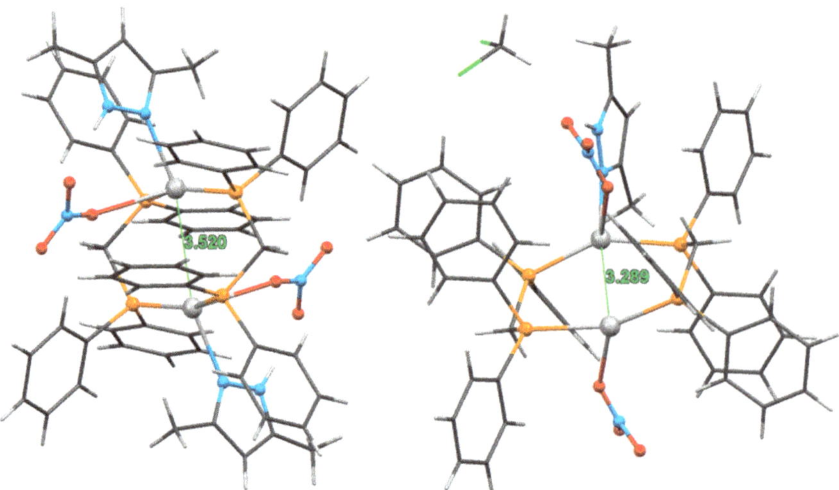

Figure 4. XRD structure of the co-crystal **2a/2b** and dichloromethane (DCM) molecule (carbon, hydrogen, and chlorine atoms are shown as sticks).

3.3. Photophysical Properties

The only pure compounds **1** and **2a**, in contrast to **2b**, were isolated. All experimental studies (UV–Vis and photoluminescence) were performed for these complexes. The theoretical investigations (TD-DFT) were performed for all possible types of complexes (**1**, **2a,** and **2b**).

The TD-DFT calculations were performed for complexes **1** and **2b** keeping their local C_i symmetry. The presence of the inversion center led to the appearance of the symmetry forbidden A_g, and the symmetry allowed A_u transitions. This is why only symmetry allowed A_u transitions are discussed further. The lowest excitations of **1** were S_1 and S_2 with energies of 4.18 and 5.01 eV (296 and 247 nm; Table S1), respectively. The S_1 had a pronounced LC^{NO3} character with a negligible impact of Ag atoms (Figure S1, Table S1). The oscillator strength (f) value of 0.0002 demonstrates that this excitation was ineffective (dark state). In contrast, the second singlet transition (5.01 eV, f = 0.1853) had a metal-centered (MC) nature (Figure 5). Upon this excitation, the electron leaves the Ag–P σ bonding orbital (impacts of Ag—40% and P—36%, Table S1) and locates on the intramolecular Ag–Ag bond, which is generated by the empty $5p_z$ orbitals of the silver cations. The vertical excitation energies for the three lowest triplets were 3.70, 4.30, and 4.60 eV (334, 287, and 269 nm), respectively. The first two triplets were of a LC^{NO3} nature, and the impact of metals and dpcym ligands appeared in the T_3 state. The T_3 had an excellent match to the S_2 state; their similarity was s_H = 0.91 and s_E = 0.95. Along with the oscillator strengths, this indicates a high probability for the excitation of **1** directly to S_2 state, $S_2 \rightarrow T_3$ interconversion, followed by the triplet emission from the T_3 state, which describes the emission observed (*vide infra*, Figure S13).

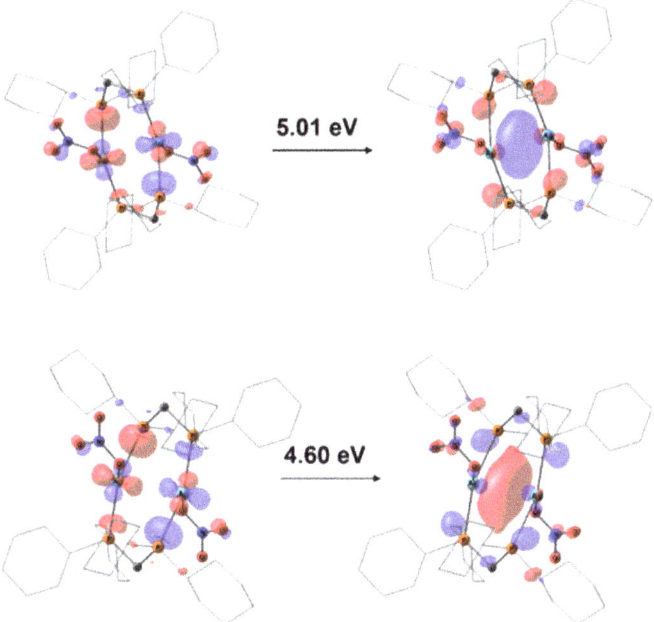

Figure 5. Highest occupied natural transition orbital (HONTO) and lowest unoccupied natural transition orbital (LUNTO) for the $S_0 \rightarrow S_2$ (**top**) and $S_0 \rightarrow T_3$ (**bottom**) transition in complex **1** as the isosurface at 0.05 a.u.

The excitation pattern of the complex $[Ag(Me_2Pz^H)(dppm)]_2(NO_3)_2$ (**2b**) differed from that observed for $[Ag(dcypm)]_2(NO_3)_2$ (**1**). The first singlet (4.16 eV, 298 nm, f = 0.2354; Figure 6) possessed a comparable impact to the hole from metals (35%) and phosphorus atoms (from dppm, 33%) with some contribution from 10% dimethylpyrazole (Figure S2, Table S2). However, the excited electron was located mainly on the organic part of the dppm ligand (64%), indicating the charge transfer character of this transition. Therefore, it could be ascribed to $(Ag^+ + P^{dppm} + N^{Pyrazole}) \rightarrow (Ph + CH_2)^{dppm}$ charge transfer. The electron density at the Ag···Ag intramolecular contact was still present in the excited S_1 state. Still, the contribution of Ag atoms to the excited state dropped by three times compared to the S_2 of complex **1**, and as a result, it could be observed only at the 0.03 a.u. isosurface (Figure S2). These data indicate the role of the dimethylpyrazole ligand in the metal-induced electronic transitions. The four lowest triplet states for **2b** were the ^{3}LC states of Ph substituents in dppm ligands, while T_5 was the $^3LC^{NO3}$ state (Figure S2). The T_6 of MLCT nature is the state coupled with the S_1 excitation, having high similarity (s_H = 0.91 and s_E = 0.90) to the $S_0 \rightarrow T_6$ transition energy 4.02 eV (308 nm) (Figure 6). Similar behavior was observed in the case of dinuclear silver pyrazolates with PCy_3 and PPh_3 ligands [35].

The TD-DFT analysis of $[Ag(dppm)]_2(Me_2Pz^H)(NO_3)_2$ (**2a**) excitations led to nearly the same pattern, except the disappearance of symmetry forbidden states, which led to doubling of the number of corresponding electronic transitions. At that, the S_1 and S_2 states both possess mixed MLCT $(Ag^+ + P^{dppm} + N^{Pyrazole}) \rightarrow (Ph + CH_2)^{dppm} + LC^{NO3}$ character with close excitation energies and oscillator strengths (4.11 eV, 301 nm, f = 0.0994 for S_1 and 4.17 eV, 297 nm, f = 0.1605 for S_2) as a result of mixing with symmetry forbidden A_G $^1LC^{NO3}$ state upon loss of symmetry (Figure 7, Table S3). Despite this mixing, the simulated excitation spectra were the same for **2a** and **2b** (Tables S2 and S3, Figure S3). For triplet transitions due to the unsymmetrical structure, the eight lowest triplets were ^{3}LC states of Ph substituents in dppm ligands, and T_9–T_{10} are $^3LC^{NO3}$ states, while T_{11} and T_{12} were MLCT states.

Due to the mixing of singlet states with LC^{NO3} transition, the discussed singlet and triplet states had a low similarity. Still, if we ignore the NO_3 group's contributions, the similarity for S_2-T_{11} was $s_H = 0.83$, $s_E = 0.79$ and for S_2–T_{12} it was $s_H = 0.84$, $s_E = 0.75$. Based on these, practically the same conclusion can be made as for **2b**. The lowest singlet excitation should lead to the high energy triplets T_{11}–T_{12} that is likely to interconvert to the lower ^3LC states of dppm′ Ph substituents, which in turn has a trend to non-emissive relaxation at room temperature [35–37], but could be stabilized at 77 K.

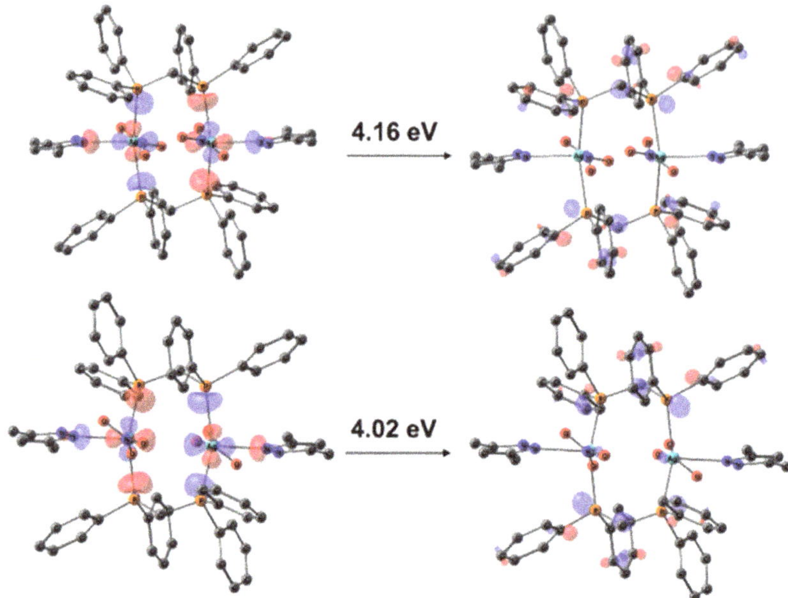

Figure 6. HONTO and LUNTO for the $S_0 \rightarrow S_1$ and $S_0 \rightarrow T_6$ in complex **2b** as the isosurface at 0.05 au.

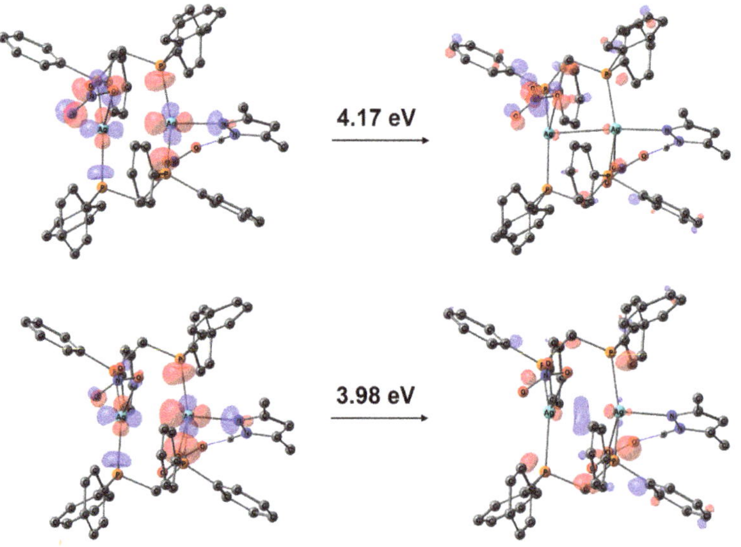

Figure 7. HONTO and LUNTO for the $S_0 \rightarrow S_2$ and $S_0 \rightarrow T_{11}$ in complex **2a** as the isosurface at 0.05 au.

The UV–Vis spectra of **1** and **2a** were measured in CH_2Cl_2 (Figure 8). Complex **1** demonstrated an intense absorbance band at 267 nm (ε = 13350 cm^{-1} M^{-}). This band can be attributed to the transition with a metal-centered (MC) character originating from intramolecular Ag(I)–Ag(I) interactions [9]. The presence of the nitrate anion in complex **1** resulted in the non-intensive tail around 300 nm, which correlated with the absorption spectrum of the aqueous solution of silver nitrate.

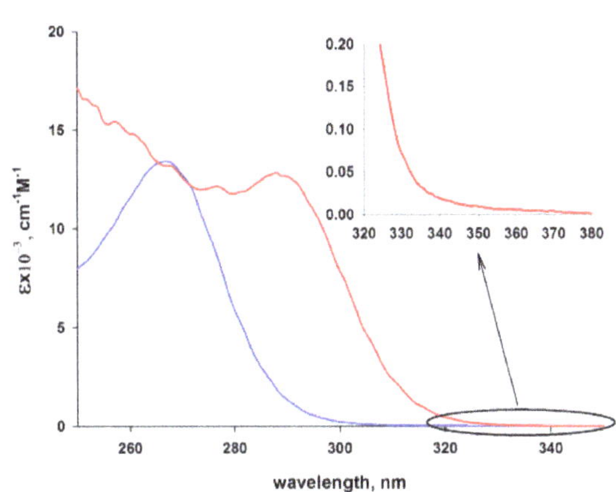

Figure 8. Absorption spectra of complexes **1** (blue) and **2a** (red) in CH_2Cl_2 solution.

UV–Vis spectrum of compound **2a** showed a broad, intense band at 288 nm (ε = 12800 cm^{-1} M^{-1}) with a long tail at ca. 335 nm, which can be assigned to the charge-transfer (MLCT or LLCT) transitions. Additionally, in the high energy region (<280 nm), weakly resolved bands with the contribution from several bands of close energies were observed. This band could be assigned only to the transitions of the $\pi \rightarrow \pi^*$ character within the Ph substituents in bisphosphine and pyrazole ligands.

The emission spectra of complexes **1** and **2a** were studied in CH_2Cl_2 solution and the solid state (Figure 9). The room temperature (RT) emission spectrum of complex **1** in the solution displayed two bands at 370 and 390 nm, respectively, which can be assigned to the emission from the electronically excited states of MC nature. The emission bands of **1** in the solid state were in the same position. Temperature decrease led to the non-significant redistribution of the solid sample band's intensities. The triplet state decays also demonstrate this mixed emissive behavior. The phosphorescence decay of **1** both in the solution and in solid state at RT and 77 K can only be fitted with bi-exponential function. In the CH_2Cl_2 solution at RT, lifetimes were 1 and 10 μs. In the solid state lifetimes, τ was practically independent of temperature being ca. 2 and 16 μs. These data demonstrate that complex **1** possessed the same photoluminescence both in solution and solid state. Complex **2a** in the acetonitrile solution possessed phosphorescence bands at 365 and 410 nm (for both τ = 8.7 μs). In the solid state at RT, complex **2a** showed an unstructured band at 380 nm of CT nature (τ = 9 μs). Interestingly, at 77 K, the presence of a new intense band at 440 nm was observed (τ = 37 μs). The position of the band observed at RT shifted from 380 to 390 nm at 77 K, but the lifetime of this emission was not dependent on the temperature (8.9 μs). The results of calculations (*vide supra*) showed that the T_1–T_8 states were of LC nature within Ph substituents. As we have shown for structurally similar silver pyrazolate/dppm complexes, these states are not emissive due to the intersection of their energy surface with the ground state, which allows effective non-radiative relaxation [35–37]. The emission observed at 298 K can be ascribed to the system relaxation from T_{11-12} (3.98–4.21 eV) and higher states of ((M + L^{P+N})L^{dppm}CT) nature. The temperature decrease led to the stabilization of triplet states of the LC^{Ph} nature and these channels possess a main influence on the emission of **2a** observed at 77 K.

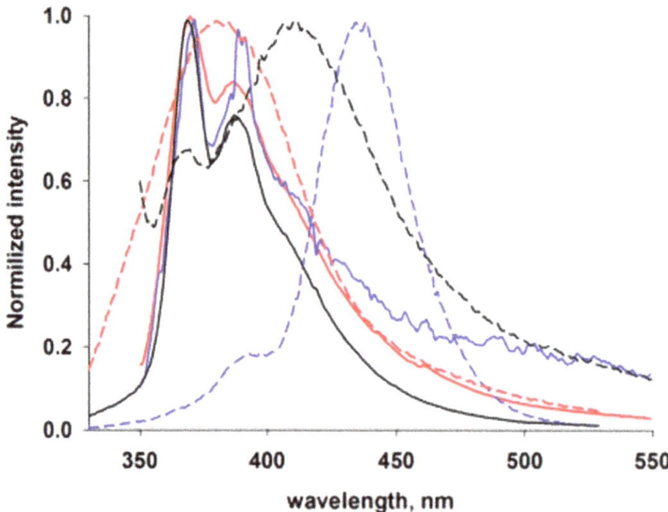

Figure 9. Normalized emission (λ_{exc} = 320 nm) spectra of complexes **1** (solid line) and **2a** (dashed line) in the solution (black) and solid state at 298 K (red) and 77 K (blue).

4. Conclusions

The reaction of silver nitrate with bisphosphines in the presence of 3,5-dimethylpyrazole gave new examples of dinuclear silver(I) complexes of different composition $[Ag(dcypm)]_2(NO_3)_2$ (**1**), $[Ag(dppm)]_2(NO_3)_2(Me_2Pz^H)$ (**2a**), and $[Ag(Me_2Pz^H)(dppm)]_2(NO_3)_2$ (**2a**). This suggests that the steric repulsion of bulky cyclohexyl substituents does not allow coordination of additional pyrazole ligand to the silver atoms in $[Ag(dcypm)]_2(NO_3)_2$. Complex **2a** was obtained by the crystallization of **2b** from a diluted solution in accordance with low solubility accompanied by the elimination of one pyrazole molecule. For all complexes, non-typical κ^1 coordination of the nitrate anion was observed, which led to a significant distortion of the central eight-member cycle $[AgPCP]_2$ in comparison to related silver bisphosphine complexes. Complexes possessed phosphorescence both in the solution and in the solid-state. The intramolecular argenthophilic interaction observed in complex **1** (d(Ag···Ag) = 2.981 Å) introduces the main excitation channel in which an electron leaves the Ag–P bond orbital and locates on the intramolecular Ag–Ag bond (metal-centered character). The coordination of dimethylpyrazole to Ag(I) in complex **2a** led to the elongation of the Ag···Ag distance (3.218 Å). Complex **2a** exhibited a phosphorescence originating from the charge-transfer from the $^3(M + L^{P+N})L^{Ph}CT$ state at room temperature. At 77 K, the emission was split into two components: $^3(M + L^{P+N})L^{Ph}CT$ and $^3LC^{Ph}$ transitions. According to the TD-DFT calculations, the coordination of dimethylpyrazole to the metal atom mainly affects the energy of transition.

Supplementary Materials: The following are available online at http://www.mdpi.com/2073-4352/10/10/881/s1, Figure S1: Natural transition orbital's of **1** as the isosurface at 0.05 a.u.; Table S1: Computed characteristics of the excited states of **1**. For analysis of fragment impacts (in %) to the ground state (hole) and excited state (electron), complex was divided to fragments of all Ag atoms, all P atoms, nitrate anions, and all atoms of dcypm ligand except P.; Figure S2: Natural transition orbital's of **2** as the isosurface at 0.05 a.u. except for S1 and T6; Table S2: Computed characteristics of the A_u symmetry excited states of **2**. For analysis of fragment impacts (in %) to the ground state (hole) and excited state (electron), the complex was divided to fragments of all Ag atoms, all P atoms, nitrate anions, pyrazolates, and all atoms of dppm ligand except P.; Figure S3: Natural transition orbital's of **2a** as the isosurface at 0.05 a.u., Table S3. Computed characteristics of the AU symmetry excited states of **2a**; Figure S4: Phosphorescence decay by delay of complex **1** at RT in CH_2Cl_2 solution (left) and solid state (right); Figure S5: Phosphorescence decay by delay of complex **1** in the solid state at 77 K; Figure S6. Phosphorescence decay by delay of complex **2a** at RT in CH_2Cl_2 solution (left) and solid state (right); Figure S7: Phosphorescence decay by delay of complex **2a** in the solid state at 77 K (measured at 390 nm); Figure S8: Phosphorescence decay by delay of complex **2a** in the solid state at 77 K (measured at 440 nm); Figure S9: The ^{1}H NMR spectrum of complex **1** in CD_2Cl_2; Figure S10: The $^{31}P\{^{1}H\}$ NMR spectrum of complex **1** in CD_2Cl_2; Figure S11: The 1H NMR spectrum of complex **2a** in CD_3CN; Figure S13: Schematic Jablonski diagram for **1**, **2a**, **2b**. Presumably active states indicated by solid lines; Figure S14: Comparison of X-ray and DFT geometry of $Ag_2P_4(NO_3)_2$ core for **1**, **2a** and **2b**. Atoms of X-ray structure is shaded, DFT structure is colored. Ag—cyan, P—orange, C—gray, O—red, N—blue; Figure S15: Excitation spectra of **1** in CH_2Cl_2 solution (left, em. on 370 and 390), and in the solid state (right, em. 370); Figure S16: Excitation spectra of **2a** in CH_2Cl_2 solution (left, em. on 410) and in the solid state (right, em. on 380); Figure S17: IR spectrum of complex **2a** in KBr pellets; Figure S18: The simulated electronic spectra of **1**, **2a**, and **2b**. The 200 lowest energy singlet transitions were considered, and full width at half maximum was set to 40; Figure S19: ^{1}H NMR spectrum of the precipitate obtained in the case of interaction silver nitrate with dppm and Me_2Pz^H, demonstrating Ag/(Me_2PzH)/dppm = 1/1/1 ratio, CD_2Cl_2; Figure S20: ^{1}H NMR spectrum of the precipitate obtained in the case of interaction silver nitrate with dppm and Me_2Pz^H, demonstrating Ag/(Me_2PzH)/dppm = 1/1/1 ratio, CD_3OD.

Author Contributions: Synthetic experiments, K.F.B.; Experiments and writing original draft, A.A.T.; Calculations and visualizations, O.A.F.; XRD studies, A.F.S.; Photoluminescence, A.A.A.; Writing review & editing, A.A.T., O.A.F., and E.S.S. All authors have read and agreed to the published version of the manuscript.

Funding: This work was financially supported by the Russian Foundation for Basic Research (RFBR grant no. 18-33-20060).

Acknowledgments: NMR and XRD measurements were performed using the equipment of the Center for Molecular Composition Studies at the INEOS RAS and were supported by the Ministry of Science and Higher Education of the Russian Federation. Photophysical properties were measured using the equipment of the Center of Physical Methods of Investigation at the IPCE RAS.

Conflicts of Interest: The authors declare that there is no conflicts of interest.

References

1. Yam, V.W.; Au, V.K.; Leung, S.Y. Light-Emitting Self-Assembled Materials Based on d(8) and d(10) Transition Metal Complexes. *Chem. Rev.* **2015**, *115*, 7589–7728. [CrossRef]
2. Schmidbaur, H.; Schier, A. Argentophilic interactions. *Angew. Chem. Int. Ed.* **2015**, *54*, 746–784. [CrossRef]
3. Dolan, N.S.; Scamp, R.J.; Yang, T.; Berry, J.F.; Schomaker, J.M. Catalyst-Controlled and Tunable, Chemoselective Silver-Catalyzed Intermolecular Nitrene Transfer: Experimental and Computational Studies. *J. Am. Chem. Soc.* **2016**, *138*, 14658–14667. [CrossRef]
4. Alderson, J.M.; Corbin, J.R.; Schomaker, J.M. Tunable, Chemo- and Site-Selective Nitrene Transfer Reactions through the Rational Design of Silver(I) Catalysts. *Acc. Chem. Res.* **2017**, *50*, 2147–2158. [CrossRef]
5. Mak, C.L.; Bostick, B.C.; Yassin, N.M.; Campbell, M.G. Argentophilic Interactions in Solution: An EXAFS Study of Silver(I) Nitrene Transfer Catalysts. *Inorg. Chem.* **2018**, *57*, 5720–5722. [CrossRef]
6. Grachova, E.V. Design of Supramolecular Cluster Compounds of Copper Subgroup Metals Based on Polydentate Phosphine Ligands. *Russ. J. Gen. Chem.* **2019**, *89*, 1102–1114. [CrossRef]
7. Ho, D.M.; Bau, R. Preparation and structural characterization of $[Ag_2(dpm)_2(NO_3)_2]$ and $[Ag_4(dpm)_4(NO_3)_2]^{2+}$ $[PF_6]^{2-}$: Conformational flexibility in the M_2P_4 core structure of bis(diphenylphosphino)methane complexes. *Inorg. Chem.* **1983**, *22*, 4073–4079. [CrossRef]
8. Che, C.-M.; Kwong, H.-L.; Yam, V.W.-W.; Cho, K.-C. Spectroscopic properties and redox chemistry of the phosphorescent excited state of $[Au_2(dppm)_2]^{2+}$[dppm = bis(diphenylphosphino)methane]. *J. Chem. Soc. Chem. Commun.* **1989**, 885–886. [CrossRef]
9. Che, C.-M.; Tse, M.-C.; Chan, M.C.W.; Cheung, K.-K.; Phillips, D.L.; Leung, K.-H. Spectroscopic Evidence for Argentophilicity in Structurally Characterized Luminescent Binuclear Silver(I) Complexes. *J. Am. Chem. Soc.* **2000**, *122*, 2464–2468. [CrossRef]
10. Lin, Y.Y.; Lai, S.W.; Che, C.M.; Fu, W.F.; Zhou, Z.Y.; Zhu, N. Structural variations and spectroscopic properties of luminescent mono- and multinuclear silver(I) and copper(I) complexes bearing phosphine and cyanide ligands. *Inorg. Chem.* **2005**, *44*, 1511–1524. [CrossRef]
11. Piché, D.; Harvey, P.D. The lowest energy excited states of the binuclear silver(I) halide complexes, $Ag_2(dmb)_2X_2$. Metal-centered or charge transfer states? *Canad. J. Chem.* **1994**, *72*, 705–713. [CrossRef]
12. Chan, J. Studies of metal binding reactions in metallothioneins by spectroscopic, molecular biology, and molecular modeling techniques. *Coord. Chem. Rev.* **2002**, *233-234*, 319–339. [CrossRef]

13. Chen, Y.; Yang, T.; Pan, H.; Yuan, Y.; Chen, L.; Liu, M.; Zhang, K.; Zhang, S.; Wu, P.; Xu, J. Photoemission mechanism of water-soluble silver nanoclusters: Ligand-to-metal-metal charge transfer vs strong coupling between surface plasmon and emitters. *J. Am. Chem. Soc.* **2014**, *136*, 1686–1689. [CrossRef]
14. Yang, T.; Dai, S.; Tan, H.; Zong, Y.; Liu, Y.; Chen, J.; Zhang, K.; Wu, P.; Zhang, S.; Xu, J.; et al. Mechanism of Photoluminescence in Ag Nanoclusters: Metal-Centered Emission versus Synergistic Effect in Ligand-Centered Emission. *J. Phys. Chem. C* **2019**, *123*, 18638–18645. [CrossRef]
15. Blake, A.J.; Donamaria, R.; Lippolis, V.; Lopez-de-Luzuriaga, J.M.; Manso, E.; Monge, M.; Olmos, M.E. Influence of crown thioether ligands in the structures and of perhalophenyl groups in the optical properties of complexes with argentoaurophilic interactions. *Inorg. Chem.* **2014**, *53*, 10471–10484. [CrossRef]
16. Donamaria, R.; Lippolis, V.; Lopez-de-Luzuriaga, J.M.; Monge, M.; Nieddu, M.; Olmos, M.E. Influence of the Number of Metallophilic Interactions and Structures on the Optical Properties of Heterometallic Au/Ag Complexes with Mixed-Donor Macrocyclic Ligands. *Inorg. Chem.* **2018**, *57*, 11099–11112. [CrossRef]
17. Ai, P.; Mauro, M.; Gourlaouen, C.; Carrara, S.; De Cola, L.; Tobon, Y.; Giovanella, U.; Botta, C.; Danopoulos, A.A.; Braunstein, P. Bonding, Luminescence, Metallophilicity in Linear Au_3 and Au_2Ag Chains Stabilized by Rigid Diphosphanyl NHC Ligands. *Inorg. Chem.* **2016**, *55*, 8527–8542. [CrossRef]
18. Sheldrick, G. SHELXT—Integrated space-group and crystal-structure determination. *Acta Crystallogr. Sect. A.* **2015**, *71*, 3–8. [CrossRef]
19. Sheldrick, G. Crystal structure refinement with SHELXL. *Acta Crystallogr. Sect. C.* **2015**, *71*, 3–8. [CrossRef]
20. Dolomanov, O.V.; Bourhis, L.J.; Gildea, R.J.; Howard, J.A.K.; Puschmann, H. OLEX2: A complete structure solution, refinement and analysis program. *J. Appl. Crystallogr.* **2009**, *42*, 339–341. [CrossRef]
21. Adamo, C.; Barone, V. Toward reliable density functional methods without adjustable parameters: The PBE0 model. *J. Chem. Phys.* **1999**, *110*, 6158–6170. [CrossRef]
22. Andrae, D.; Häußermann, U.; Dolg, M.; Stoll, H.; Preuß, H. Energy-adjustedab initio pseudopotentials for the second and third row transition elements. *Theor. Chim. Acta* **1990**, *77*, 123–141. [CrossRef]
23. Weigend, F.; Ahlrichs, R. Balanced basis sets of split valence, triple zeta valence and quadruple zeta valence quality for H to Rn: Design and assessment of accuracy. *Phys. Chem. Chem. Phys.* **2005**, *7*, 3297–3305. [CrossRef]
24. Frisch, M.J.; Trucks, G.W.; Schlegel, H.B.; Scuseria, G.E.; Robb, M.A.; Cheeseman, J.R.; Scalmani, G.; Barone, V.; Mennucci, B.; Petersson, G.A.; et al. *Gaussian 09, Revision B.01*; Gaussian Inc.: Wallingford, CT, USA, 2009.
25. Lu, T.; Chen, F. Multiwfn: A multifunctional wavefunction analyzer. *J. Comput. Chem.* **2012**, *33*, 580–592. [CrossRef]
26. Chen, R.; Tang, Y.; Wan, Y.; Chen, T.; Zheng, C.; Qi, Y.; Cheng, Y.; Huang, W. Promoting Singlet/triplet Exciton Transformation in Organic Optoelectronic Molecules: Role of Excited State Transition Configuration. *Sci. Rep.* **2017**, *7*, 6225. [CrossRef]
27. Das, S.; Sharma, S.; Singh, H.B.; Butcher, R.J. Metallophilic Mercuraazamacrocycles Derived from Bis{6-formyl-(2,3,4-trimethoxy)phenyl}mercury: Reactivity with d10 and d8 Metal Ions. *Eur. J. Inorg. Chem.* **2018**, *2018*, 4093–4105. [CrossRef]
28. Titov, A.A.; Filippov, O.A.; Smol'yakov, A.F.; Averin, A.A.; Shubina, E.S. Synthesis, structures and luminescence of multinuclear silver(I) pyrazolate adducts with 1,10-phenanthroline derivatives. *Dalton Trans.* **2019**, *48*, 8410–8417. [CrossRef]
29. Emashova, S.K.; Titov, A.A.; Filippov, O.A.; Smol'yakov, A.F.; Titova, E.M.; Epstein, L.M.; Shubina, E.S. Luminescent Ag^I Complexes with 2,2′-Bipyridine Derivatives Featuring [Ag-$(CF_3)_2$Pyrazolate]$_4$ Units. *Eur. J. Inorg. Chem.* **2019**, *2019*, 4855–4861. [CrossRef]
30. Desiraju, G.R. Hydrogen bridges in crystal engineering: Interactions without borders. *Acc. Chem. Res.* **2002**, *35*, 565–573. [CrossRef]
31. Steiner, T. The Hydrogen Bond in the Solid State. *Angew. Chem. Int. Ed.* **2002**, *41*, 48–76. [CrossRef]
32. Tiekink, E.R.T. Bis-μ-[methylenebis(diphenylphosphine)]-dinitratodisilver(I) dichloroform solvate. *Acta Crystallogr. Sect. C.* **1990**, *46*, 235–238. [CrossRef]
33. Cui, Y.-Z.; Yuan, Y.; Li, Z.-F.; Liu, M.; Jin, Q.-H.; Jiang, N.; Cui, L.-N.; Gao, S. From ring, chain to network: Synthesis, characterization, luminescent properties of silver(I) complexes constructed by diphosphine ligands and various N-donor ligands. *Polyhedron* **2016**, *112*, 118–129. [CrossRef]

34. Qiu, Q.-M.; Huang, X.; Zhao, Y.-H.; Liu, M.; Jin, Q.-H.; Li, Z.-F.; Zhang, Z.-W.; Zhang, C.-L.; Meng, Q.-X. Synthesis, structure and terahertz spectra of six Ag(I) complexes of bis(diphenylphosphino)methane with 4,4′-bipyridine and its derivations. *Polyhedron* **2014**, *83*, 16–23. [CrossRef]
35. Titov, A.A.; Filippov, O.A.; Smol'yakov, A.F.; Baranova, K.F.; Titova, E.M.; Averin, A.A.; Shubina, E.S. Dinuclear Cu(I) and Ag(I) Pyrazolates Supported with Tertiary Phosphines: Synthesis, Structures, and Photophysical Properties. *Eur. J. Inorg. Chem.* **2019**, *2019*, 821–827. [CrossRef]
36. Titov, A.A.; Filippov, O.A.; Smol'yakov, A.F.; Averin, A.A.; Shubina, E.S. Copper(I) complex with BINAP and 3,5-dimethylpyrazole: Synthesis and photoluminescent properties. *Mendeleev Commun.* **2019**, *29*, 570–572. [CrossRef]
37. Titov, A.A.; Filippov, O.A.; Smol'yakov, A.F.; Godovikov, I.A.; Shakirova, J.R.; Tunik, S.P.; Podkorytov, I.S.; Shubina, E.S. Luminescent Complexes of the Trinuclear Silver(I) and Copper(I) Pyrazolates Supported with Bis(diphenylphosphino)methane. *Inorg. Chem.* **2019**, *58*, 8645–8656. [CrossRef]

MDPI

Article

A Novel Fishbone-Like Lead(II) Supramolecular Polymer: Synthesis, Characterization, and Application for Producing Nano Metal Oxide

Younes Hanifehpour [1,*], Babak Mirtamizdoust [2], Sang Woo Joo [3,*], Majid Sadeghi-Roodsari [2] and Mehdi Abdolmaleki [1]

1 Department of Chemistry, Faculty of Science, Sayyed Jamaleddin Asadabadi University, Asadabad 6541861841, Iran; m.abdolmaleki@sjau.ac.ir

2 Department of Chemistry, Faculty of Basic Sciences, University of Qom, P.O. Box 37185-359, Qom, Iran; babakm.tamizdoust@gmail.com (B.M.); chemmj94@gmail.com (M.S.-R.)

3 School of Mechanical Engineering, WCU Nano Research Center, Yeungnam University, Gyeongsan 712-749, Korea

* Correspondence: Hanifehpour@sjau.ac.ir (Y.H.); swjoo@yu.ac.kr (S.W.J.)

Abstract: The nanorods of $[Pb(L)Br_2]_n$ (**1**) (L = 1,2-bis (pyridin-3-ylmethylene)hydrazine) underwent ultrasound irradiation and were synthesized as a novel three-dimensional fishbone-like Pb(II)–organic coordination supramolecular compound. The morphology and nanostructure of the synthesized compound were determined through SEM, FTIR, elemental analyses, and XRD. Compound **1** was structurally characterized by single-crystal X-ray diffraction and revealed six-coordinated Pb (II) ions bonded to two N atoms from two L ligands and four bromine anions, forming a one-dimensional fishbone-like coordination polymer, which extended into a 3D supramolecular structure through weak intermolecular interactions. The bulk thermal stability of compound **1** was examined using thermogravimetric analysis (TGA). Moreover, PbO nanoparticles with sizes of 40–80 nm were obtained through the thermolysis of **1** at 180 °C using oleic acid as a surfactant.

Keywords: nanorods; Pb(II); ultrasonic irradiation; nano metal oxide; crystal structure

Citation: Hanifehpour, Y.; Mirtamizdoust, B.; Joo, S.W.; Sadeghi-Roodsari, M.; Abdolmaleki, M. A Novel Fishbone-Like Lead(II) Supramolecular Polymer: Synthesis, Characterization, and Application for Producing Nano Metal Oxide. *Crystals* **2021**, *11*, 335. https://doi.org/10.3390/cryst11040335

Academic Editor: Alexander S. Novikov

Received: 1 March 2021
Accepted: 24 March 2021
Published: 26 March 2021

1. Introduction

Coordination compounds or metal–organic hybrid materials have remarkably attracted huge attention during the past twenty years. Inorganic–organic hybrid supramolecular materials include the self-assembly of an organic ligand with proper metal ions and functional groups with special directionality and functionality [1–5]. Structural blocks are created based on coordinating interactions and other poor non-covalent inter-molecular forces, including hydrogen bonds and pi stacking, which have key roles in their stability. The attraction to these compounds comes from the effects of basic structural chemistry, and their probable applications in areas including luminescence, catalysis, nonlinear optics, magnetism and molecular sensing and adsorption, which do not exist for mononuclear compounds [6–8]. Lowering the size of nano-scaled coordination supramolecular compounds increases the surface area. Preparing any types of nano-scaled coordination supramolecular compounds is therefore the main path toward the technological application of these novel materials [9].

Pb(II) frameworks have attracted interest because of their large ion radius, variable coordination number, and possible occurrence of a stereochemically active lone pair of $6s^2$ outer electrons with a novel network of topologies and interesting properties [10–12]. According to the hard–soft acid–base theory, the intermediate coordination ability of Pb(II) means that it can flexibly coordinate small nitrogen or oxygen atoms as well as large sulfur atoms [13]. Investigation of the stereo-chemical activity of valence shell lone electron pairs in the polymeric and supramolecular compounds may be more interesting. The spontaneous aggregation of several bridging ligands causes the gap to disappear, and the coordination of Pb (II) assumes a less common holodirected arrangement [14].

The present study introduces an eco-friendly and simple nanocrystal synthesis of a three-dimensional coordination supramolecular compound via ultrasonic irradiation. Organic synthesis was performed, and nanomaterials prepared using sonochemical techniques [15]. Further studies are still required for clarifying the application of the Ultrasound technique (*UST*) to constructing coordination supramolecular compounds. The present study investigated the fast synthesis of $[Pb(L)Br_2]_n$ (**1**) [L = 1,2-bis (pyridin-3-ylmethylene)hydrazine] as the nanocrystals of a one-dimensional fishbone-like Pb(II) coordination polymer. According to the findings obtained, Ultrasound technique (*UST*) synthesis is an efficient, low cost, simple and eco-friendly method for coordination supramolecular nanocompounds compared to conventional synthetic methods such as the solvent diffusion method and solvothermal and hydrothermal approaches [16].

Ultrasound technique (*UST*) leads to high-energy chemistry by processing acoustic cavitation, including bubbles, expansion, and their implosive collapse within a liquid intermediate [17]. UST results in alternative compressive and expansive acoustic waves to make and oscillate bubbles by irradiating liquids. Bubble can invade and consequently collapse fast, i.e., with cooling and heating rates of over 10^{10} Ks^{-1}. The huge energy concentration obtained over the collapse leads to a pressure of about one thousand bar and a local temperature of approximately 5000 K. The energy distribution to the surroundings during collapse and after bounce causes the hot spot gas temperature to swiftly reach the temperature of the surrounding area [18,19]. Ultrasonic irradiation can therefore be used to assist various chemical reactions progress at room temperature, and even certain reactions formerly difficult to actualize using conventional approaches [20–23]. The present study employed oleic acid as a surfactant and performed thermolysis of **1** at 180 °C to synthesize and structurally characterize a novel nano lead(II) metal–organic supramolecular compound with 1,2-bis (pyridine-3-ylmethylene)hydrazine ligand (Scheme 1) and simply synthesize PbO nanoparticles.

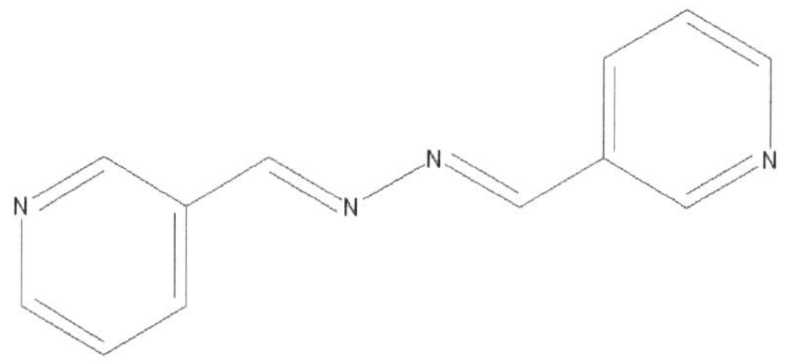

N N N N

Scheme 1. 1,2-bis (pyridine-3-ylmethylene)hydrazine ligand.

2. Experiments

2.1. Measurements and Materials

Chemicals were obtained from Sigma-Aldrich (Seoul, South Korea) and used as received without further purification. Elemental analyses of the samples were carried out using a Vario Microanalyzer. FTIR was carried out by employing Bruker Vector 22 and KBr disks at 4000–400 cm^{-1}. Thermogravimetric analysis (TGA) was conducted under argon flow at 20–600 degrees Celsius and a heating rate of 3 °C/min using LABSYS evo (SETARAM).

X-ray powder diffraction (XRD) measurements were carried out using an X'pert diffractometer (Panalytical) with monochromatized Cu-Kα radiation. In addition, XRD patterns were simulated in Mercury to acquire single-crystal information [24]. The morphology of the nanostructured compound was determined by scanning electron microscopy (SEM) (S-4200, Hitachi, Japan) and TEM (JEM-2200FS, JEOL). A 40 kHz Sonicator_3000 multiwave ultrasonic generator made by Misonix Inc. in the UST equipped with a titanium oscillator (horn) and a converter/transducer 12.5 mm in diameter was utilized to perform ultrasound. The minimum power output of the generator was 600 W for one hour at ambient temperature. The chemical composition and status of the product were evaluated by X-ray photoelectron spectroscopy (XPS) (K-ALPHA, UK).

2.2. Preparation of 1,2-bis (pyridine-3-ylmethylene)hydrazine (L)

A mixture of 5 mmol (0.45 mL, 0.45 g) of hydrazine hydrate (35 wt.% in water) and a solution of 10 mmol (1.07 g) of nicotinaldehyde in 100 mL of methanol was refluxed for one day. Afterwards, the obtained yellow solids were filtered and then rinsed in methanol and dried in a vacuum (yield: 65%, 0.98 g), m.p. 140 °C, FTIR (KBr) = ν_{max} (698, 706, 817, 863, 1021, 1089, 1185, 1225, 1304, 1417, 1482, 1586, 1625, 2929) cm^{-1}. H NMR ($CDCl_3$, ppm): 8.66 (s, 2H), 8.70 (d, 2H), 8.97 (s, 2H), 7.41 (d, 2H), 8.22 (d, 2H).

2.3. Preparation of Nanostructure and Single Crystal $[Pb(L)Br_2]_n$ (1)

A total of 30 mL of a 0.1 M solution of $PbBr_2$ was added to 30 mL of a 0.1 M solution of the (*L*) ligand, and sonication performed with a high-density 40 kHz, 600 W UST probe to make the nanostructured $[Pb(L)Br_2]_n$ (**1**). The obtained precipitate was filtered and rinsed in water and then dried in the air at the decomposition point: 330 °C. According to the analyses, C: 25.0%, H: 2.0% and N: 10.0% and C: 24.9%, H: 1.5% and N: 9.7% were obtained for $C_{12}H_{10}Br_2N_4Pb$. The selected FTIR bands (cm^{-1}) were as follows: 635*s*, 750*s*, 864*m*, 1045*m*, 1418*s*, 1482*m*, 1590*s*, 1620*s*, 3020*w*.

An appropriate single crystal of $[Pb(L)Br_2]_n$ (**1**) determining the X-ray structure was made and isolated by inserting 1 mmol of *L* into an arm of a tube with branches and 1 mmol of $PbBr_2$ to the other arm [25]. Having filled both the arms with methanol and sealing the tube, the arm containing the ligand was submerged in a 60 °C oil bath. Yellow crystals (decomposition point of 333 °C) were precipitated for 9 days on the other arm that was placed at room temperature. Afterwards, the crystals were filtered, rinsed in water, and dried in air with a yield of 70% (0.402 g). The results obtained were as follows: H: 2.0%, C: 25.5% and N: 10.0%. Moreover, C: 24.9%, H: 1.5% and N: 9.7% were obtained for $C_{12}H_{10}Br_2N_4Pb$. The selected FTIR bands (cm^{-1}) were as follows: 634*s*, 750*s*, 864*m*, 1045*m*, 1418*s*, 1482*m*, 1590*s*, 1620*s*, 3024*w*.

2.4. Preparation of Nanoparticles of Pb(II) Oxide

After dissolving 0.1 mmol of $[Pb(L)Br_2]_n$ (**1**) in 10 mL of oleic acid, the light-yellow solution obtained was degassed for forty-five minutes (with stirring and slow heating) and then heated for one hour at 180 °C, ultimately yielding a black precipitate. After adding a great amount of ethanol and a little toluene to the solution, lead(II) oxide nanoparticles were centrifuged and thereby isolated. The final solid was rinsed in ethanol and dried at room temperature (yield: 43%, 0.01 g).

2.5. X-ray Crystallography

A 0.65 × 0.15 × 0.09 mm^3 yellow compound crystal was placed over glass fiber using epoxy adhesives. An X-ray diffractometer (50 kV, 30 mA) with graphite monochromated Mo K_α radiation (λ = 0.71073 Å) was used over 2θ = 4.56–50.08° to collect data. No significant data were missing from the data collected. The data were processed in the crystal structure analysis module of the Bruker AXS [26]. The modules of AXS utilized included SADABS (Bruker) for absorption correction, SHELXS-97 (Sheldrick) and XPREP (Bruker) for structure solution, SHELXTL (Sheldrick) for molecular graphics and publication materials, SHELXL-97 (Sheldrick) for structure refinement, APEX2 (Bruker) for data collection, and SAINT (Bruker) for cell refinement and data reduction. Furthermore, the method proposed by Waber and Cromer was used to obtain scattering factors for neutral atoms [27]. The crystal was of the monoclinic space group $P2_1/c$ given the systematic deficiency, E statistic, and effective structure refining. Furthermore, the structure was solved by utilizing direct techniques. For the data obtained from the compound, full-matrix least squares refinement was performed by minimizing the function of $\sum w(F_c^2 - F_0^2)^2$. Moreover, non-hydrogen atoms were anisotropically refined. The location of the hydrogen atoms was geometrically determined with 0.98Å (CH_3) and C–H = 0.95 (aromatic) and refined with Uiso(H) = −1.2 UeqC as riding atoms. Tables 1 and 2 present the crystallographic data, the bond lengths, and the angles.

Table 1. Crystal data and structure refinement for $[Pb(L)Br_2]_n$ (**1**).

Empirical Formula	$C_{12}H_{10}Br_2N_4Pb$
Formula weight	555.23
Temperature/K	293(2)
Crystal system	monoclinic
Space group	$P2_1/c$
a/Å	4.005(7)
b/Å	9.169(13)
c/Å	34.27(4)
α/°	90.00
β/°	92.08(4)
γ/°	90.00
Volume/Å^3	1258(3)
Z	4
ρ_{calc}mg/mm^3	2.922
μ/mm^{-1}	19.751
F(000)	1000.0
Crystal size/mm^3	0.65 × 0.15 × 0.09
2Θ range for data collection	4.76 to 50.08°
Index ranges	$-4 \leq h \leq 3, -10 \leq k \leq 10, -37 \leq l \leq 40$
Reflections collected	6637
Independent reflections	1789[R(int) = 0.0834]
Data/restraints/parameters	1789/0/160
Goodness-of-fit on F^2	1.101
Final R indexes [I > = 2σ (I)]	$R_1 = 0.1166$, $wR_2 = 0.2949$
Final R indexes (all data)	$R_1 = 0.1255$, $wR_2 = 0.3013$
Largest diff. peak/hole/e Å^{-3}	5.58/−3.65

Table 2. Selected bond lengths (Å) and angles (°) for $[Pb(L)Br_2]_n$ (**1**).

Pb1—N2	2.67 (2)	Pb1—Br1	2.992 (4)
Pb1—N2^{i}	2.67 (2)	Pb1—Br1^{iii}	2.975 (4)
Pb1—Br1^{ii}	2.975 (4)	Pb1—Br1^{i}	2.992 (4)
N2—Pb1—N2^{i}	180.000 (12)	N2—Pb1—Br1^{ii}	91.2 (6)
N2—Pb1—Br1	84.4 (5)	Br1^{iii}—Pb1—Br1^{i}	84.32 (13)
N2^{i}—Pb1—Br1	95.6 (5)	Br1^{ii}—Pb1—Br1^{i}	95.68 (13)
N2^{i}—Pb1—Br1^{iii}	91.2 (6)	Br1^{iii}—Pb1—Br1	95.68 (13)
N2—Pb1—Br1^{iii}	88.8 (6)	Br1^{ii}—Pb1—Br1	84.32 (13)
N2—Pb1—Br1^{i}	95.6 (5)	Br1^{i}—Pb1—Br1	180.00 (9)
N2^{i}—Pb1—Br1^{i}	84.4 (5)	Br1^{iii}—Pb1—Br1^{ii}	180.00 (9)
N2^{i}—Pb1—Br1^{ii}	88.8 (6)		

Symmetry code(s): (i) $-x+2, -y+1, -z+1$; (ii) $-x+1, -y+1, -z+1$; (iii) $x+1, y, z$.

3. Results and Discussion

$[Pb(L)Br_2]_n$ (**1**) was obtained as a novel one-dimensional fishbone coordination polymer from the reaction of lead(II) bromide with 1,2-bis (pyridine-3-ylmethylene)hydrazine (L). The nanostructure of the compound was determined in an aqueous solution through UST irradiation. The suitable single crystals of compound I for X-ray crystallography were obtained using the heat gradient of an aqueous solution based on the branched-tube method [28].

According to Figure 1, the bulk material developed using the branched-tube approach could not be differentiated from the nanostructured material obtained using the sonochemical approach in terms of FTIR spectrum and the results of elemental analysis. The FTIR spectrum of the single crystal materials and nanostructured material showed the representative absorption bands of the "L" ligand. Furthermore, the aromatic C–H absorption of hydrogen atoms caused a relatively weak band at approximately 3021 cm^{-1}. The aromatic ring vibration of the "L" ligand was also observed at frequencies of 1418–1590 cm^{-1}.

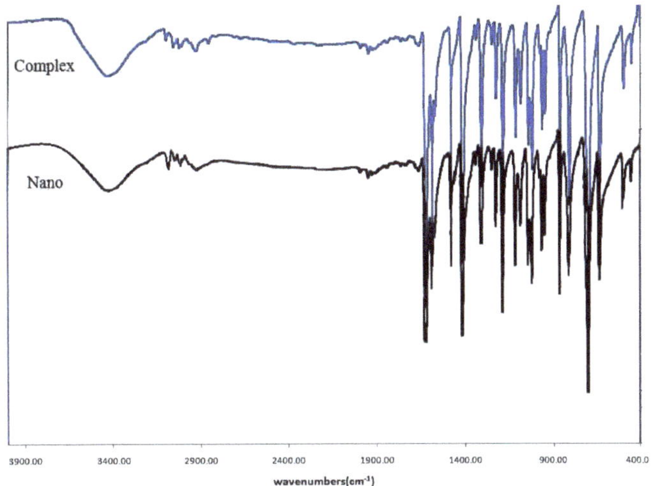

Figure 1. FT-IR spectrum of crystal and nano structure of **1**.

Tables 1 and 2 show the results of single-crystal XRD for the structure of compound I. According to the single-crystal XRD in Figure 2, the structure of $[Pb(L)Br_2]_n$ crystallized in a monoclinic system with the space group of $P2_1/c$ resembled that of solid 1D fishbone metal–organic coordination polymers. Figure 2a shows the structure of the asymmetrical unit of **1** and an atom numbering arrangement selected. A Pb atom is coordinated by two N atoms of the *L* ligand with the same Pb–N distance of 2.670 Å and four bridged bromine anions with a Pb–Br distance of 2.992 (2.972, 2.992 and 2.972 Å). Figure 2b shows a coordination number of six for lead(II) atom in two modes with a PbN_2Br_4 donor set. The adjusted Pb····Pb distances belonging to the chain were 4.001 Å. This order demonstrates a symmetry in the coordination geometry surrounding metal ions in a holodirected fashion. Given the labile interactions, the architecture of one-dimensional chains was allowed to interact with neighboring chains, and the structure was also allowed to be extended into three-dimensional supramolecular metal-organic coordination polymers, as shown in Figure 2d.

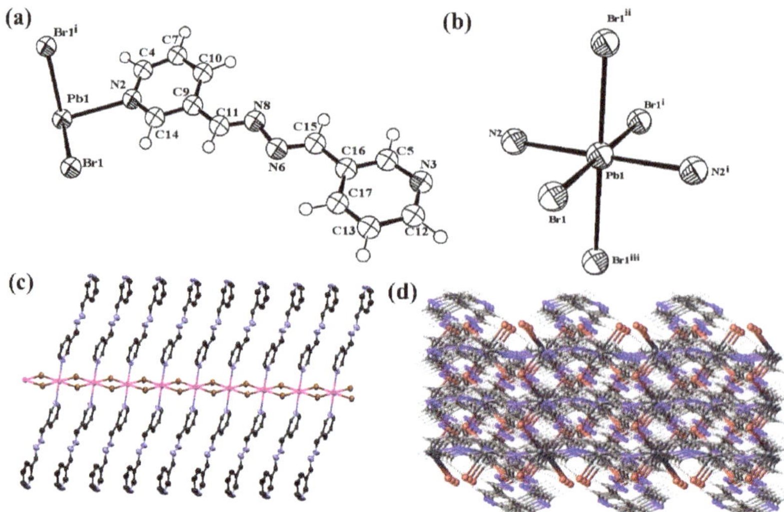

Figure 2. (**a**) Molecular structure of asymmetrical unit (**1**). (**b**) Coordination sphere of Pb(II) in **1**. (**c**) Fragment of the coordination polymer showing the 1D fish bone metal–organic coordination polymer. (**d**) From 1D architecture to supramolecular 3D coordination polymer via labile interactions (such as aromatic and π–π interaction, H-bonding).

Figure 3a shows the XRD pattern of compound **1** simulated from the single-crystal X-ray data, and Figure 3b shows the experimental XRD pattern of compound **1** prepared by using the sonochemical process. Except for slight changes in 2θ, the experimental data were consistent with the simulated XRD patterns, suggesting that the compound obtained as a nano-structure using the sonochemical method was the same as that derived through single-crystal XRD. The major peak enlargmenet showed the nano-dimension of the particles. According to $D = 0.891\lambda/\beta\cos\theta$ as the Scherrer equation, the mean size of the grain was estimated at 39 nm, with D representing the mean grain size, $\lambda = 0.15405$ nm as the X-ray wavelength, and θ and β as the diffraction and full-width angles, respectively, at 50% of a peak.

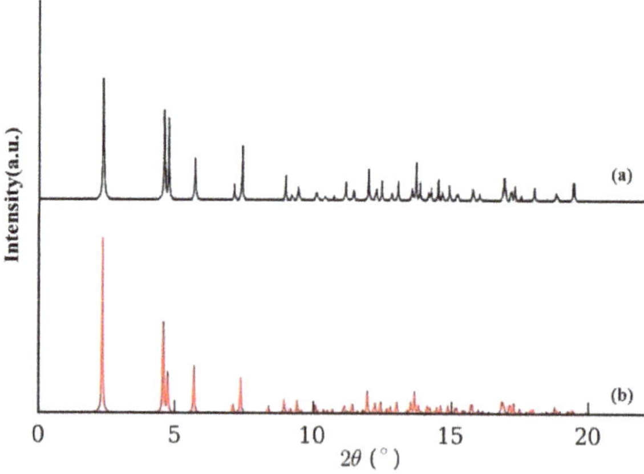

Figure 3. X-ray Powder Diffraction Pattern (XRPD) of compound **1** (**a**) computed from single-crystal X-ray data and (**b**) from the nanorods.

The crystalline substance was provided with $[Pb(L)Br_2]_n$ (**1**) through the reaction of 1,2-bis (pyridin-3-ylmethylene)hydrazine (L) with lead(II) bromide. In addition, the size and morphology of compound I prepared by the UST technique were analyzed with SEM. Figure 4 shows the SEM of compound I acquired by using a 600 W ultrasonic generator at a $[Pb^{2+}] = [L^-] = 0.1$ M concentration of primary reagents. Figure 4 also shows the nanorod morphology with a diameter of 45–140 nm for compound I. Further assessments are required for clarifying the formation of these structures. Packing the system at molecular levels, however, morphologically affects the nanostructured compound, suggesting coordination-encouraged creation over the morphology of the nanostructure [29–32].

Figure 4. SEM photographs of $[Pb(L)Br_2]_n$ (**1**) nanorods.

Figure 5 shows the results of the TGA of the single crystals and nanostructures performed at 25–600 °C in argon to investigate the thermal stability of compound 1. For single crystals of $[Pb(L)Br_2]_n$, after removing water molecules from the coordinated molecules and moisture at 80–110 °C, compound 1 was stable at up to 160 °C. Decomposition at 160–425 °C resulted in a mass loss of about 47.8%. In the case of nanostructured $[Pb(L)Br_2]_n$, the decomposition started at a lower temperature, around 145 °C, with the similar weight loss. This may be attributed to different crystallographic surfaces of both samples exposed to a thermal gradient [33]. The thermal stability of $[Pb(L)Br_2]_n$ is much lower than our previous reports on lead (II) coordination polymer, which may be attributable to differences in the structures of ligands coordinated to lead [34,35]. XPS findings showed that the solid residue at about 600 °C was PbO.

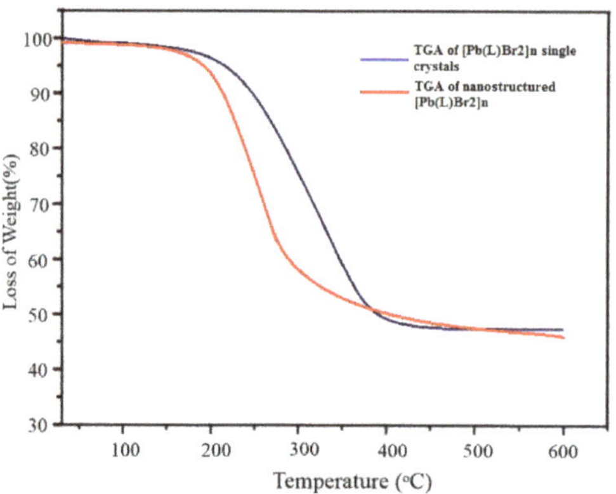

Figure 5. TGA plot of $[Pb(L)Br_2]_n$ (**1**) in single crystal shape and nanostructured mode.

The final products of the decomposition of compound 1 based on their XRD arrangements as per Figure 6a matched the standard design of tetragonal PbO with S.G = P4/nmm, a = 3.947 Å, Z = 2, c = 4.988 Å and JCPDS (Joint Committee on Powder Diffraction Standards) card file No: 85-0711. Figure 6b displays the XPS results of nano PbO. The two asymmetric peaks at 142.28 eV and 137.48 eV were attributed to the transitions of $4f_{5/2}$ and $4f_{7/2}$ from Pb^{2+} ions in PbO, respectively, indicating that Pb^{2+} in the product was the same as PbO in shape. The O1 s spectrum showed a peak at 531.1 eV, which was estimated at 530.9 eV of O1 s in PbO in the literature. Figure 7 shows the SEM and TEM results of the residue obtained from the thermolysis of **1** at 180 °C by employing oleic acid as a surfactant, suggesting the regular shape of Pb(II) oxide nanoparticles. The significant broadening of the peaks indicates that the particles have nanometer dimensions. The average grain size D was estimated as 58 nm using the Scherrer formula, $D = 0.891\lambda/\beta cos\theta$, where λ is the X-ray wavelength (0.15405 nm) and θ and β are the diffraction angle and full-width at half maximum of an observed peak, respectively [36].

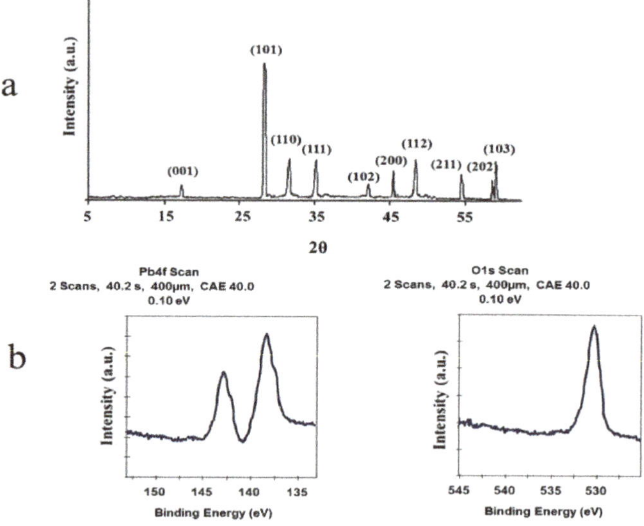

Figure 6. (**a**) XRD pattern and (**b**) XPS spectra of PbO nanoparticles prepared by thermolysis of compound **1**.

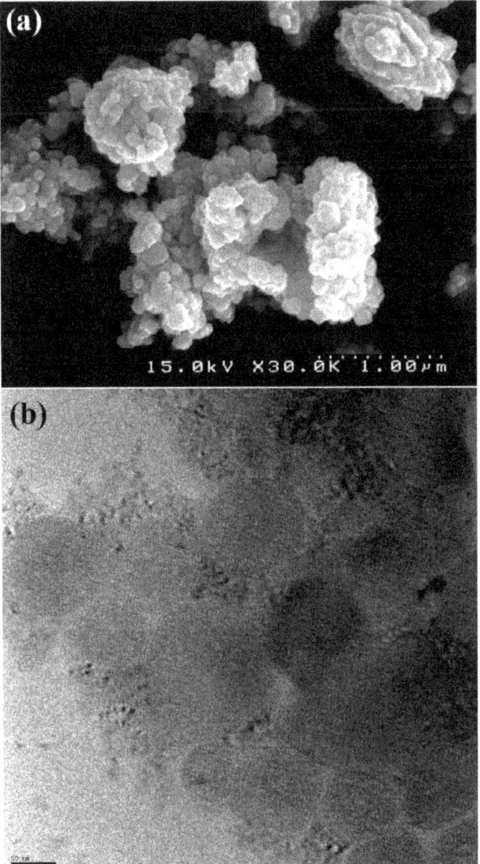

Figure 7. (**a**) SEM and (**b**) TEM images of PbO nano-structures produced by the thermolysis of nano-structures of compound **1**.

4. Conclusions

In this study, a sonochemical method was used to synthesize nano-structures of a new 1D fishbone coordination polymer of divalent lead with 1,2-bis (pyridin-3-ylmethylene)-hydrazine (L). Single-crystal XRD was performed to structurally determine compound **1**. A three-dimensional supramolecular compound included in the crystal structure of compound I demonstrated a coordination number of six in lead (II) ions. The complex takes the form of a 1D metal–organic system in solid state. As a result of several labile interactions with neighboring chains, the 1D chain extended into a 3D supramolecular coordination polymer. The sonochemically made compound **1** had nanorod morphology with diameter of 38–43 nm. Moreover, the calcination of compound I through the thermolysis of **1** at 180 °C by using oleic acid as a surfactant yielded a uniform PbO nanoparticle.

Author Contributions: Y.H., supervision, writing—review and editing; B.M., writing—review and editing; S.W.J., project administration and editing; M.S.-R., writing—original draft preparation; M.A., formal analysis. All authors have read and agreed to the published version of the manuscript.

Funding: This work was funded by the grant NRF-2019R1A5A8080290 of the National Research Foundation of Korea.

Institutional Review Board Statement: Not applicable.

Informed Consent Statement: Not applicable.

Data Availability Statement: Cambridge Crystallographic Data Centre provided the crystallographic information for the structures as supplementary publication CCDC-1567103 for $[Pb(L)Br_2]_n$ (**1**). Duplicates of these data may be acquired by contacting deposit@ccdc.cam.ac.uk by email or through official correspondence with CCDC, 12 Union Road, Cambridge CB2 1EZ, United Kingdom, Fax: +441223336033.

Acknowledgments: The authors thank Sayyed Jamaledin Asadabadi University and Qom University for their support.

Conflicts of Interest: The authors declare no conflict of interest.

References

1. Chakrabarty, R.; Mukherjee, P.S.; Stang, P.J. Supramolecular coordination: Self-assembly of finite two-and three-dimensional ensembles. *Chem. Rev.* **2011**, *111*, 6810–6918. [CrossRef]
2. Hu, M.L.; Morsali, A.; Aboutorabi, L. Lead (II) carboxylate supramolecular compounds: Coordination modes, structures and nano-structures aspects. *Coord. Chem. Rev.* **2011**, *255*, 2821–2859. [CrossRef]
3. Leong, W.L.; Vittal, J.J. One-dimensional coordination polymers: Complexity and diversity in structures, properties, and applications. *Chem. Rev.* **2011**, *111*, 688–764. [CrossRef] [PubMed]
4. Akhbari, K.; Morsali, A. Thallium (I) supramolecular compounds: Structural and properties consideration. *Coord. Chem. Rev.* **2010**, *254*, 1977–2006. [CrossRef]
5. Gharib, M.; Safarifard, V.; Morsali, A. Ultrasound assisted synthesis of amide functionalized metal-organic framework for nitroaromatic sensing. *Ultrason. Sonochem.* **2018**, *42*, 112–118. [CrossRef] [PubMed]
6. Robin, A.Y.; Fromm, K.M. Coordination polymer networks with O-and N-donors: What they are, why and how they are made. *Coord. Chem. Rev.* **2006**, *250*, 2127–2157. [CrossRef]
7. Kitagawa, S.; Kitaura, R.; Noro, S. Functional porous coordination polymers. *Angew. Chem. Int. Ed.* **2004**, *43*, 2334–2375. [CrossRef] [PubMed]
8. Tanhaei, M.; Mahjoub, A.R.; Safarifard, V. Sonochemical synthesis of amide-functionalized metal-organic framework/graphene oxide nanocomposite for the adsorption of methylene blue from aqueous solution. *Ultrason. Sonochem.* **2018**, *41*, 189–195. [CrossRef] [PubMed]
9. Uemura, T.; Kitagawa, S. Nanocrystals of coordination polymers. *Chem. Lett.* **2005**, *34*, 132–137. [CrossRef]
10. Shimoni-Livny, L.; Glusker, J.P.; Bock, C.W. Lone pair functionality in divalent lead compounds. *Inorg. Chem.* **1998**, *37*, 1853–1867. [CrossRef]
11. Engelhardt, L.M.; Furphy, B.M.; Harrowfield, J.M.; Patrick, J.M.; White, A.H. 1: 1 Adducts of lead (II) thiocyanate with 1, 10-phenanthroline and 2,2′,6′,2″-terpyridine. *Inorg. Chem.* **1989**, *28*, 1410–1413. [CrossRef]
12. Akhbari, K.; Morsali, A.; Retailleau, P. Effect of two sonochemical procedures on achieving to different morphologies of lead (II) coordination polymer nano-structures. *Ultrason. Sonochem.* **2013**, *20*, 1428–1435. [CrossRef] [PubMed]
13. Platas-Iglesias, C.; Esteban-Gómez, D.; Enríquez-Pérez, T.; Avecilla, F.; de Blas, A.; Rodriguez-Blas, T. Lead (II) thiocyanate complexes with bibracchial lariat ethers: An X-ray and DFT study. *Inorg. Chem.* **2005**, *44*, 2224–2233. [CrossRef] [PubMed]
14. Taheri, S.; Mojidov, A.; Morsali, A. Two-dimensional Holodirected Lead (II) Coordination Polymer, [Pb (μ-bpp)(μ-SCN) 2] n (bpp= 1, 3-di (4-pyridyl) propane). *Z. Anorg. Allg. Chem.* **2007**, *633*, 1949–1951. [CrossRef]
15. Gedanken, A. Using sonochemistry for the fabrication of nanomaterials. *Ultrason. Sonochem.* **2004**, *11*, 47–55. [CrossRef] [PubMed]
16. Bang, J.H.; Suslick, K.S. Applications of ultrasound to the synthesis of nanostructured materials. *Adv. Mater.* **2010**, *22*, 1039–1059. [CrossRef] [PubMed]
17. Son, W.J.; Kim, J.; Kim, J.; Ahn, W.S. Sonochemical synthesis of MOF-5. *Chem. Commun.* **2008**, *47*, 6336–6338. [CrossRef] [PubMed]
18. Suslick, K.S. The sonochemical hot spot. *J. Acoust. Soc. Am.* **1991**, *89*, 1885–1886. [CrossRef]
19. Safarifard, V.; Morsali, A. Facile preparation of nanocubes zinc-based metal-organic framework by an ultrasound-assisted synthesis method; precursor for the fabrication of zinc oxide octahedral nanostructures. *Ultrason. Sonochem.* **2018**, *40*, 921–928. [CrossRef] [PubMed]
20. Kojima, Y.; Koda, S.; Nomura, H. Effects of sample volume and frequency on ultrasonic power in solutions on sonication. *Jpn. J. Appl. Phys.* **1998**, *37*, 2992. [CrossRef]
21. Koda, S.; Kimura, T.; Kondo, T.; Mitome, H. A standard method to calibrate sonochemical efficiency of an individual reaction system. *Ultrason. Sonochem.* **2003**, *10*, 149–156. [CrossRef]
22. Cravotto, G.; Boffa, L.; Mantegna, S.; Perego, P.; Avogadro, M.; Cintas, P. Improved extraction of vegetable oils under high-intensity ultrasound and/or microwaves. *Ultrason. Sonochem.* **2008**, *15*, 898–902. [CrossRef] [PubMed]
23. Safarifard, V.; Morsali, A. Applications of ultrasound to the synthesis of nanoscale metal–organic coordination polymers. *Coord. Chem. Rev.* **2015**, *292*, 1–14. [CrossRef]
24. *Mercury 1.4.1*; Copyright Cambridge Crystallographic Data Centre: Cambridge, UK, 2001–2005.

25. Harrowfield, J.M.; Miyamae, H.; Skelton, B.W.; Soudi, A.A.; White, A.H. Lewis-base adducts of lead (II) compounds. XX. synthesis and structure of the 1: 1 adduct of pyridine with lead (II) thiocyanate. *Aust. J. Chem.* **1996**, *49*, 1165–1169. [CrossRef]
26. SHELXTL, N.T. *Crystal Structure Analysis Package, Version 6.14*; Bruker AXS: Madison, WI, USA, 2000.
27. Cromer, D.T.; Waber, J.T. *International Tables for X-ray Crystallography*; Table 2.2 A; Kynoch Press: Birmingham, UK, 1974; Volume 4.
28. Morsali, A.; Masoomi, M.Y. Structures and properties of mercury (II) coordination polymers. *Coord. Chem. Rev.* **2009**, *253*, 1882–1905. [CrossRef]
29. Fard, M.J.S.; Morsali, A. Sonochemical Synthesis of New Nano-Belt One-Dimensional Double-Chain Lead (II) Coordination Polymer; As Precursor for Preparation of PbBr (OH) Nano-Structure. *J. Inorg. Organomet. Polym. Mater.* **2010**, *20*, 727–732. [CrossRef]
30. Ranjbar, Z.R.; Morsali, A.; Retailleau, P. Thermolysis preparation of zinc (II) oxide nanoparticles from a new micro-rods one-dimensional zinc (II) coordination polymer synthesized by ultrasonic method. *Inorg. Chim. Acta* **2011**, *376*, 486–491. [CrossRef]
31. Ranjbar, Z.R.; Morsali, A. Preparation of Zinc (II) Oxide Nanoparticles from a New Nano-Size Coordination Polymer Constructed from a Polypyridyl Amidic Ligand; Spectroscopic, Photoluminescence and Thermal Analysis Studies. *J. Inorg. Organomet. Polym. Mater.* **2011**, *21*, 421–430. [CrossRef]
32. Masoomi, M.Y.; Mahmoudi, G.; Morsali, A. Sonochemical syntheses and characterization of new nanorod crystals of mercury (II) metal–organic polymer generated from polyimine ligands. *J. Coord. Chem.* **2010**, *63*, 1186–1193. [CrossRef]
33. Marzbanrad, E.; Rivers, G.; Peng, P.; Zhao, B.; Zhou, N.Y. How morphology and surface crystal texture affect thermal stability of a metallic nanoparticle: The case of silver nanobelts and pentagonal silver nanowires. *Phys. Chem. Chem. Phys.* **2015**, *17*, 315–324. [CrossRef] [PubMed]
34. Hanifehpour, Y.; Mirtamizdoust, B.; Ahmadi, H.; Wang, R.; Joo, S.W. Ultrasonic-assisted synthesis, characterizing the structure and DFT calculation of a new Pb(II)-chloride metal-ligand coordination polymer as a precursor for preparation of α-PbO nanoparticles. *J. Mol. Struct.* **2021**, *1124*, 129031. [CrossRef]
35. Hanifehpour, Y.; Mirtamizdoust, B.; Ahmadi, H.; Wang, R.; Joo, S.W. Sonochemical synthesis, crystal structure, and DFT calculation of an innovative nanosized Pb(II)-azido metal–organic coordination polymer as a precursor for preparation of PbO nanorod. *Chem. Pap.* **2020**, *74*, 3651–3660. [CrossRef]
36. Patterson, A.L. The Scherrer formula for X-ray particle size determination. *Phys. Rev.* **1939**, *56*, 978. [CrossRef]

Article

Ultrasound-Assisted Synthesis and Crystal Structure of Novel 2D Cd (II) Metal–Organic Coordination Polymer with Nitrite End Stop Ligand as a Precursor for Preparation of CdO Nanoparticles

Younes Hanifehpour [1,*], Jaber Dadashi [2] and Babak Mirtamizdoust [2]

[1] Department of Chemistry, Sayyed Jamaleddin Asadabadi University, Asadabad 6541861841, Iran
[2] Department of Chemistry, Faculty of Science, University of Qom, P.O. Box 37185-359, Qom, Iran; jabber_dadashi_95@yahoo.com (J.D.); babakm.tamizdoust@gmail.com (B.M.)
* Correspondence: younes.hanifehpour@gmail.com or Hanifehpour@sjau.ac.ir

Abstract: In the present research, a sonochemical approach was applied to prepare new cadmium(II) coordination 2D polymer, $[Cd(L)(NO_2)_2]_n$ (L = 1,2-bis(1-(pyridin-3-yl)ethylidene)hydrazine) and structurally characterized with various spectroscopic techniques including XRD, elemental analysis, SEM, and IR spectroscopy. The coordination number of cadmium (II) ions is seven (CdN_2O_5) by two nitrogen atoms from two organic Schiff base ligand and five oxygen of nitrite anions. The 2D sheet structures ended by nitrite anions and the nitrite anion displayed the end-stop role. The comprehensive system showed a three-dimensional structure with several weak interactions. The high-intensity ultrasound is regarded as an easy, environmentally-friendly, and flexible synthetic instrument for the compounds of coordination. CdO NPs was obtained by thermolysing 1 at 180 °C with oleic acid (as a surfactant). Further, the size and morphology of the produced CdO nanoparticles were investigated through SEM.

Keywords: crystal structure; coordination polymer; sonochemical route; X-ray crystallography

Citation: Hanifehpour, Y.; Dadashi, J.; Mirtamizdoust, B. Ultrasound-Assisted Synthesis and Crystal Structure of Novel 2D Cd (II) Metal–Organic Coordination Polymer with Nitrite End Stop Ligand as a Precursor for Preparation of CdO Nanoparticles. *Crystals* **2021**, *11*, 197. https://doi.org/10.3390/cryst11020197

Academic Editor: F. Christopher Pigge

Received: 23 December 2020
Accepted: 12 February 2021
Published: 17 February 2021

1. Introduction

Recently, the development of metal–organic coordination systems has become increasingly attractive for researchers [1–3]. Ascetic architecture, luminescence, topologic diversity, magnetic, storage molecular, adsorption, covering, various preparation manners, and other interesting possible usage are among the factors attracting attention and results in various examinations leading to many research works in the area of coordination polymers and novel group of material chemistry [4–6].

Schiff bases that are a significant series of organic ligands have the capability of coordination by metal ions for forming stable chelates through the N atom of the azo-group as well as other donors, like $-NH_2$ and –OH. Schiff bases have advantages of satisfactory biological activity, simple synthetic pathways, high coordination potency, and photochemical features, and they include pharmaceuticals (e.g., antibiotics, anti-inflammatory drugs, and antivirals) and also photoelectric and dye materials [7–12]. Moreover, because of the variety of structure, potential features, and preparative accessibility, metal complexes with Schiff base ligands are considered a significant stereochemical model in metal coordination chemistry. The common coordination mode of these ligands is different and very interesting [13]. Because of the relatively simple yet robust synthetic procedure, Schiff base formation and complexation have formed the basis of a number of elegant reported undergraduate experiments that explore various aspects of Schiff base chemistry, from the formation of organometallic complexes to combinatorial synthesis to spectral analysis. In this experiment, Schiff base ligands are used to demonstrate the effect of metal ions on metal–ligand stoichiometry [14].

Metal–organic coordination systems can be used as templates in order to obtain the desired nano-materials. Via selecting proper metal–organic compounds precursors with special morphologies and under suitable experimental conditions, acquiring the desired morphologies will become possible. Nano metal–organic coordination compounds precursors may also be used to obtain nano-materials with better and more useful properties [15–18]. Depending on the conditions, metal–organic coordination compounds precursor may lead to the preparation of metals, metal oxides, metal sulfides, or other nano-sized materials. With this method, nano-materials can be obtained in different morphologies from different metal–organic coordination systems precursors.

A small number of the transition metals' characteristic attributes are demonstrated by Cadmium and group 12 elements because they do not partially fill d or f electron shells. Cadmium, like zinc, has a preference for a + 2 oxidation state in the majority of its complexes. However, the more important point is that these elements with spherical d^{10} configuration are the facilitators of different coordination geometries, allowing forming of 1D, 2D, and 3D coordination compounds [19–21].

The focus of the present study is the preparation and design of metal–organic coordination systems, and the synthesis of some nano metal–organic coordination polymers have been recently reported by our research team [22–26]. The next phase of our team in this regard is an investigation of the behaviors of Cadmium (II) in terms of coordination by 2-bis(1-(pyridin-3-yl)ethylidene)hydrazine) Schiff-base ligand in the present work.

The near similar works on these compounds were reported in previous literature with various composite and applicant (see [27]). Especially, the ligand's structural chemistry is attractive due to its multifunctional coordination state which four aromatic and aliphatic nitrogen can be coordinated to metals (Scheme 1). A simple sonochemical preparation process is described in this work for a nanostructure for this coordination system and its usage in the construction of Cadmium oxide (CdO) NPs.

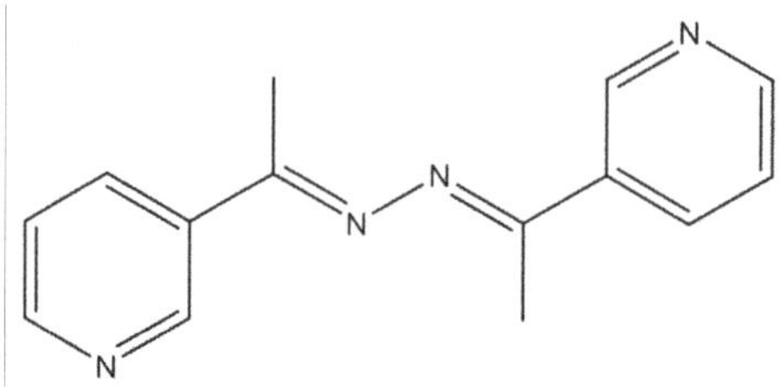

Scheme 1. Schematic of 2-bis (1-(pyridin-3-yl)ethylidene)hydrazine) Schiff-base ligand.

2. Experimental

2.1. Instruments and Materials

All chemical compounds were bought from the Aldrich and Merck (Germany and Merck Company, Darmstadt, Germany) chemical companies. A Bruker Vector 22 FT-IR (Richmond Scientific, Chorley, UK) spectrometer with KBr disks in the area of 400–4000 cm^{-1} was used for acquiring IR spectroscopy. An Electrothermal 9000 (Keison Products, Chelmsford, UK) apparatus determine the melting points. A multiwave ultrasonic generator (Sonicator_3000; Misonix, Inc., Farmingdale, NY, USA) was applied. X'pert diffractometer (produced by Panalytical, Panalytical, Malvern, UK) monochromatized Cu K_{α} radiation was employed for performing measurements of PXRD. Mercury 2.4 was used for preparing simulated PXRD powder patterns on the basis of single-crystal data [28]. Elemental analyses were carried out using a linked ISIS-300, Oxford EDS (Oxford Instruments, Abingdon, UK) (energy dispersion spectroscopy) detector. Chemical states and chemical compositions of the CdO nanoparticles were carried out through X-ray photoelectron spectroscopy (XPS) (Model: K-ALPHA, Thermo Fisher, Waltham, MA, USA).

The Scherrer formula was employed for estimating the crystallite size of chosen samples. Followed by gold coating Au, the samples' morphology was assessed using an

SEM (Hitachi, Japan). Using an XcaliburTM diffractometer with a Sapphire2 CCD detector and Mo Kα radiation (monochromator Enhance, Oxford Diffraction Ltd., Abingdon, UK) and ω-scan rotation methods at 150 K, diffraction data were collected for the single crystal. The CrysAlis program package (Oxford Diffraction Ltd., UK) was utilized for data reduction and gathering and refinement of cell parameters [29].

Multi-scan absorption modification with the CrysAlis (Rigaku, Tokyo, Japan) was used for the [Cd(L)(NO2)2]n data. With a full matrix weighted least-squares method (SHELX-2014) (Free Software Foundation, Boston, MA, US) with the $w = 1/[\sigma^2(F_o)^2 + (0.035P)^2]$ weight, where $P = (F_o^2 + 2F_c^2)/3$, we solved the structure with direct processes by SHELXS and adjusted on all F^2 data anisotropically. Mercury 2.4 was applied for creating molecular exhibits. Crystal data and arrangement modification for 1 are provided in Table 1, and the designated bond lengths and angles can be observed in Table 2.

Table 1. Crystal data and structure refinements of complex.

Chemical Formula	$C_{14}H_{14}CdN_6O_4$
M_r	442.7
Crystal system, space group	Orthorhombic, $P2_12_12_1$
Temperature (K)	120
a, b, c (Å)	4.5208 (1), 16.2510 (2), 22.0539 (4)
V (Å^3)	1620.25 (5)
Z	4
Radiation type	Cu $K\alpha$
μ (mm^{-1})	11.12
Crystal size (mm)	0.38 × 0.10 × 0.03
Diffractometer	Goniometer Xcalibur, detector: Atlas (Gemini ultra Cu)
Absorption correction	Analytical *CrysAlis PRO*, Agilent Technologies (2010), Analytical numeric absorption correction using a multifaceted crystal model
T_{min}, T_{max}	0.181, 0.723
No. of measured, idenpendent and observes [1 > 2σ(1) reflections	20706, 2895, 2742
R_{int}	0.039
$(\sin\theta/\lambda)_{max}$(Å^{-1})	0.598
Refinement $R[F^2 > 2\sigma(F^2)]$, $wR(F^2)$, S	0.020, 0.057, 1.13
No. of reflections	2895
No. of parameters	227
H-atom treatment	H-atom parameters constrained
$\Delta\rangle_{max}$, $\Delta\rangle_{min}$ (e Å$^{-3)}$	0.30, −0.37

Table 2. Chosen bond lengths [Å] & angles [°] for 1.

1.234 (4)	**O2—N5**	**2.390 (2)**	**Cd1—O1**
1.252 (5)	O3—N6	2.368(2)	Cd1—O3
1.255 (5)	O4—N6	2.470(2)	Cd1—O2
1.335 (5)	N1—C1	2.389(2)	Cd1—O4
1.340 (5)	N1—C5	2.328 (2)	Cd1—O1^i
1.397 (5)	N2—N3	2.358 (2)	Cd1—N1^i
1.291 (5)	N2—C6	2.356 (2)	Cd1—N4
1.346 (5)	N4—C12	1.288 (5)	N3—C8
120.2 (2)	Cd1^{ii}—N1—C5	1.339 (5)	N4—C11
154.44 (9)	N1^i—Cd1—N4	87.27 (11)	O1—Cd1—N1^i
121.9 (2)	Cd1^{ii}—N1—C1	89.09 (10)	O1—Cd1—N4

Symmetry code(s): (i) $-x, y + 1/2, -z + 1/2$; (ii) $-x, y-1/2, -z + 1/2$.

2.2. *Preparation of 1,2-Bis(1-(pyridin-3-yl)ethylidene)hydrazine) Ligand*

20 mmol (1.9 mL, 2.42 g) of 1-(pyridin-3-yl)ethan-1-one was dissolved in 50 mL of ethanol, then 10 mmol (0.9 mL, 0.91 g) of hydrazine hydrate (35 wt % solution in water), softened in 20 mL of ethanol, was mixed in a drop-wise manner; subsequently, two acetic

acid drops were mixed with it too. The stirring of the reaction mix was done for 24 h at ambient temperature. Finally, the yellow sediments were filtered and air-dried (yield: 77%, 2.56 g product) (Comparable with previous synthesized [30,31]).

FT-IR (KBr, cm^{-1}): 758(m), 823(w), 563(m), 1290(s), 1365(s), 1415(s), 1485(s), 1558(s), 1600(s).

Elemental analysis calculated for $C_{14}H_{14}N_4$: C: 70.57, H: 5.92, N: 23.51%.

Experimental: C: 70.10, H: 6.00, N: 24.00%.

2.3. Synthesis of Nano-Structure of $[Cd(L)(NO_2)_2]_n$

For synthesizing, the nanostructure of $[Cd(L)(NO_2)_2]_n$ a solution of cadmium(II) nitrate tetrahydrate (15 mL of a 0.1 M) in water was put in a high-density ultrasonic probe operating at 20 kHz by the highest power output of 60 W. 15 mL of a 0.1 M solution of the Schiff base ligand and 30 mL of a 0.1 M solution of sodium nitrite were added into this solution in a drop-wise manner. The resulting yellow precipitates were filtered, rinsed with H_2O and MeOH, and dried at 50 °C (0.69 g product/yield: 65% decompose point: 210 °C).

Elemental analysis; calculated for $C_{14}H_{14}CdN_6O_4$: C: 37.98, H: 3.19, N: 18.98%.

Experimental: C: 38.00, H: 4.10, N: 20.00.

2.4. Synthesis of Isolate Single Crystal of $[Cd(L)(NO_2)_2]_n$

For isolating $[Cd(L)(NO_2)_2]_n$ single crystals, Schiff base ligand (1 mmol/0.24 g), sodium nitrite (2 mmol/0.14 g) and cadmium(II) nitrate tetrahydrate (1 mmol/0.31 g) were put in the primary arm of a branched tube for heating (Scheme 2) [32].

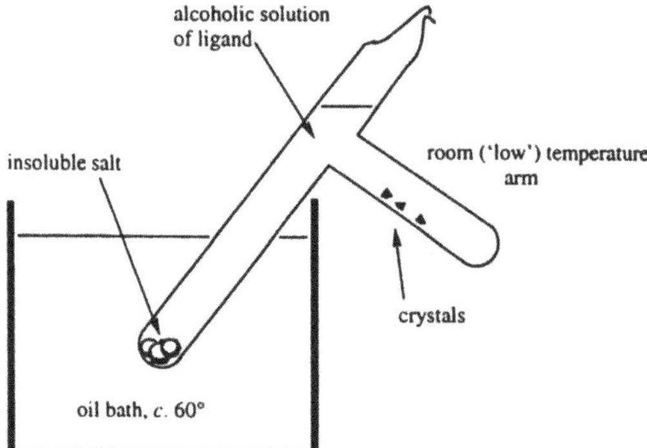

Scheme 2. Brached tube method [32].

Both arms received methanol carefully. The tube was sealed, and the arm containing ligand was submerged in an oil bath at 60 °C; however, the other arm was stored at ambient temperature. Following two weeks, yellow crystals were precipitated in the colder arm for X-ray structure determination that was filtered, rinsed by acetone and ether, and being dried with exposure to air. (0.59 g product/Yield 89%—decompose point: 210 °C) Elemental analysis; calculated for C14H14CdN6O4: C: 37.98, H: 3.19, N: 18.98%. Experimental: C: 38.00, H: 4.00, N: 20.00

3. Synthesis of CdO Nanoparticles

The precursor of [Cd(L)(NO2)2]n (**1**) (0.02 mmol) was dissolved instantly in 1.0 mL of oleic acid, which formed a light yellow solution. This solution was degassed for 20 min and then heated to 180 °C for 2 h. At the end of the reaction, an orange precipitate was formed. A small amount of toluene and a large excess of EtOH were added to the reaction solution, and cadmium (II) oxide nanoparticles were separated by centrifugation. The light brown solids were washed with EtOH and dried under an air atmosphere (0.01 g product/yield: 58%).

4. Results and Discussion

The reaction of 1,2-bis(1-(pyridin-3-yl)ethylidene)hydrazine) Schiff base ligand with a mixture of sodium nitrite and cadmium nitrate, the novel Cd (II) metal–organic coordination polymer $[Cd(L)(NO_2)_2]_n$ was obtained. Using ultrasonic irradiation in a methanolic solution, the compound's nanostructure was achieved; then, the single crystalline substance was attained through application of the heat gradients to the reagents solution or branched tube route. Scheme 3 presents the two various routes utilized for preparing $[Cd(L)(NO_2)_2]_n$ materials.

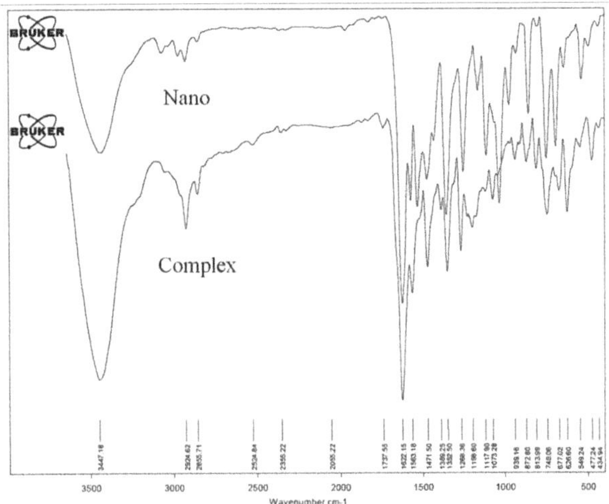

Figure 1. FT–IR spectra of the complex and nano-structures of coordination polymer.

Schiff base ligand + $NaNO_2$ + $Cd(NO_3)_2$

$MeOH/H_2O$, Ultrasound, 1h → nano structure of $[Cd(L)(NO_2)_2]n$

MeOH, Conventional heating, 14 d → single crystal of $[Cd(L)(NO_2)_2]n$

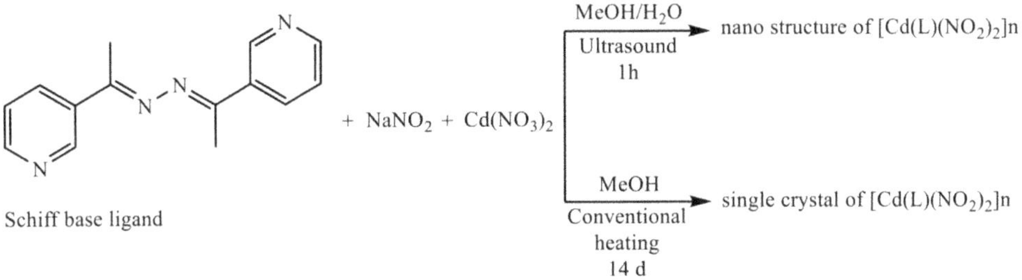

Scheme 3. The synthetic techniques and the produced materials. The FT-IR spectra of synthesized crystal and nano-structure of coordination polymer ($[Cd(L)(NO_2)_2]_n$) indicate the characteristic absorption bands for organic Schiff base ligand. The weak band near 3000–3100 cm^{-1} is assigned to aromatic C–H groups' stretching vibrations. The C=N tensile band is a sharp and strong peak in the 1622 area. The absorption band in the 1400–1600 cm^{-1} range is attributed to the Schiff base ligand's aromatic ring vibrations. There is a good match between FT-IR spectra of the complex and nano-structure in all areas (Figure 1).

The surface morphology of the synthesized nano coordination compound was investigated through FE-SEM (field emission scanning microscopy) analysis. The morphology of the compound prepared by the sonochemical method is block-like aggregates (Figure 2). SEM assessments illustrate that achieved nanoparticles are of various sizes and have different shapes. Figure 3 showed the Simulated XPRD pattern of $[Cd(L)(NO_2)_2]_n$ in comparison with nanostructure XRD pattern, Acceptable matches with slight differences in 2θ were observed between the simulated and experimental XPRD patterns (Figure 3). This indicates that the synthesis obtained by the sonochemical method as nanoparticles is identical to that obtained by crystallography. The significant broadening of the peaks indicates that the particles are of nanometer dimensions. As it has been estimated by the Sherrer formula ($D = 0.891\lambda/\beta cos \cdot \theta$, where D is the average grain size, λ the X-ray wavelength (0.15405 nm), and θ and β the diffraction angle and full-width at half maximum of an observed peak, respectively) [33], the obtained value is D = 83 nm.

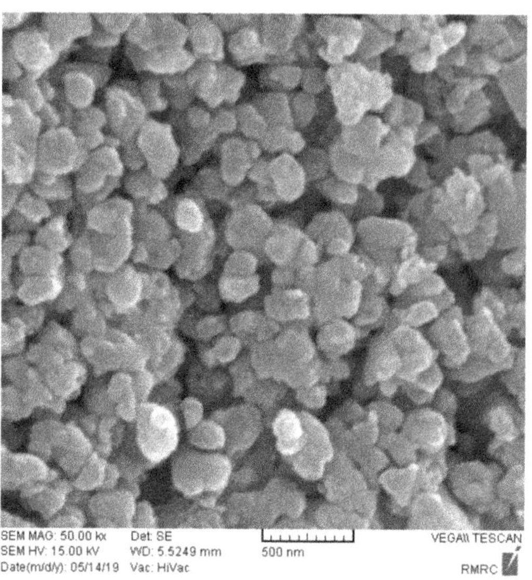

Figure 2. SEM images of the $[Cd(L)(NO_2)_2]_n$.

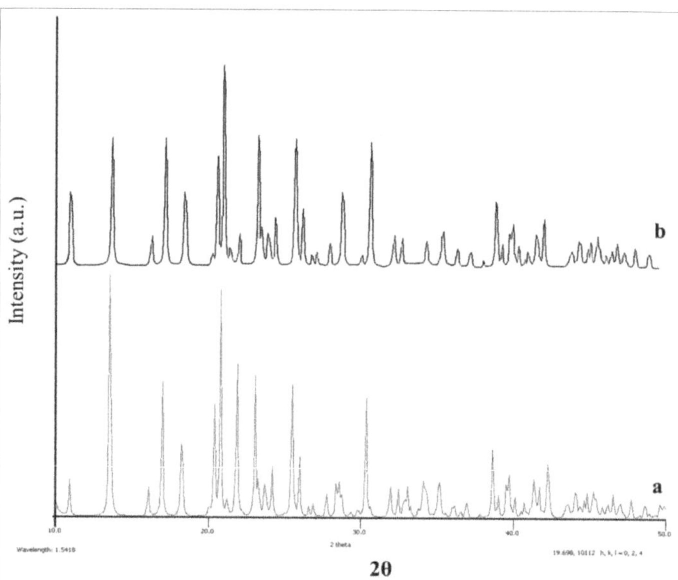

Figure 3. The PXRD patterns of compound $[Cd(L)(NO_2)_2]_n$ (**1**): (**a**) Simulated from single-crystal X-ray data and (**b**) nanostructure of (**1**) prepared by sonochemical method. Based on the crystal structure of **1**, the composition and stereochemistry of a building block $[Cd(L)(NO_2)_2]_n$) were disclosed. A view of the molecular structure is shown in Figure 4.

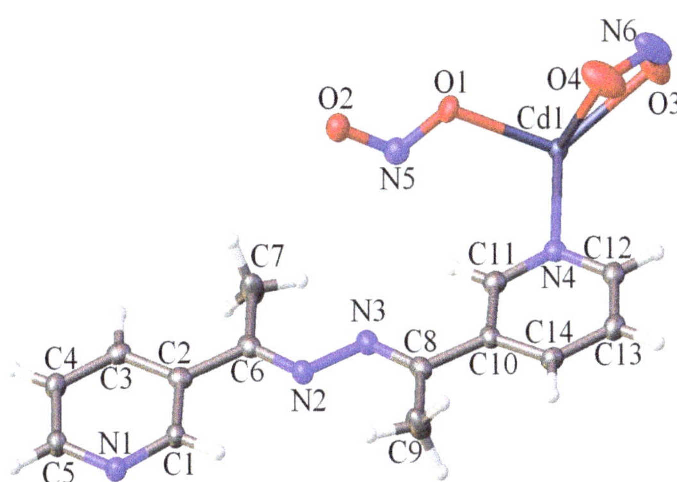

Figure 4. Asymmetric unit of the $[Cd(L)(NO_2)_2]_n$.

The determination of the structure of coordination polymer $[Cd(L)(NO_2)_2]_n$ displayed that the complex crystallizes in the Orthorhombic system with space group $P2_12_12_1$, taking the form of a two-dimensional polymer in the solid-state (Figure 5). The Cd (II) atom is coordinated with two nitrogen atoms of two Schiff base ligands with the Cd–N distances of 2.356 (2), 2.358 (2) Å, five oxygen atoms of three nitrite anions. In addition distance of Cd–O bonds in this complex are around 2.368–2.247 Å. Therefore, the coordination number of the Cd (II) atom is seven, with an orthorhombic system with space group $P2_12_12_1$ (Figures 4 and 5) (See Supplementary Materials for further details). The 2D sheet structures ended with nitrite anions. In this structure, the nitrite anion showed the end-stop role and by connecting directly to the metal, does not allow the system to expand to the third dimension.

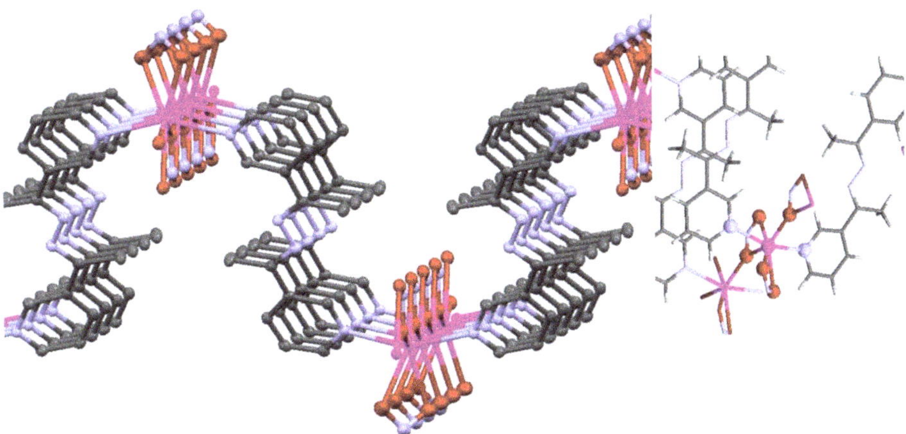

Figure 5. The coordination polymer fragment showing the 2D arrangement of complex (Left), The coordination sphere of Cd(II) (Right).

The structure of other compounds reported from this ligand to Mn, Ag, Co, Zn, Bi, and Hg metals with varying coordination numbers for metals and the same coordination method for ligands is comparable [34–38].

There are many weak interactions exist in the structure; (like short contact involving C-H groups and aromatic interactions) (See Table 3. For example, there are pi–pi stacking interactions between the N1C5C4C3C3C1 plans about 3.770 Å and between the N4C12C13C14C10C11 plans about 3.474 Å. With several weak interactions, the 2D chains interact with neighbors, and the system extended to 3-dimensional structures (Figure 6 and Table 3).

Table 3. Selected non-covalent contacts in the crystal structure of **1** [Å and °].

D-H...A	d(D-H)	d(H...A)	d(D...A)	<(DHA)
C(13)-H(13)...O(4)	0.96	2.525	3.237(4)	130.9
C(13)-H(7)...C(12)	0.96	2.708	2.789(4)	155.5
C(4)-H(4)...O(3)	0.95	2.572	3.442(5)	151.2
C(1)-H(1)...N(5)	0.96	2.514	3.328(4)	142.6
C(5)-H(5)...O(3)	0.95	2.641	3.674(5)	145.7

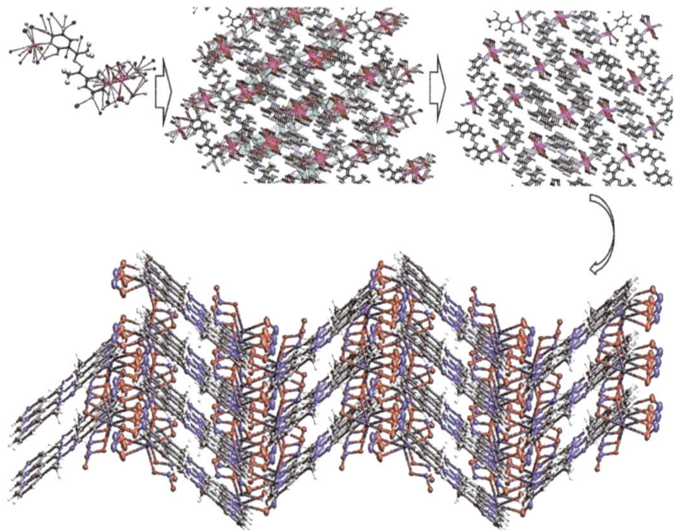

Figure 6. From di nuclear complexes to 3D networks.

The decomposition of precursor $[Cd(L)(NO_2)_2]_n$ in oleic acid functioning as a surfactant under an air atmosphere at 180 °C yields nano-powder of cadmium (II) oxide. The PXRD pattern of as-synthesized CdO nanopowder is seen in Figure 7. All the diffraction peaks are matched well with the cubic crystal structure of pure CdO with respect to their positions as found in JCPDS card No.05-0640.

Figure 8 displays the energy-dispersive X-ray spectroscopy (EDS) of the CdO nanoparticles and confirms the existence of the elements Cd/O and the formation of CdO as expected. The atomic percentage composition of the elements Cd and O were observed to be equal to 50.79 and 49.21, respectively.

Figure 9a displays the SEM image of as-prepared CdO nanoparticles. The obtained CdO has the regular nanoparticle shape with diameters of around 30–80 nm according to the size distribution graph of the product which was determined using Nahamin microstructure distance measurement software (Iran). (Figure 9b).

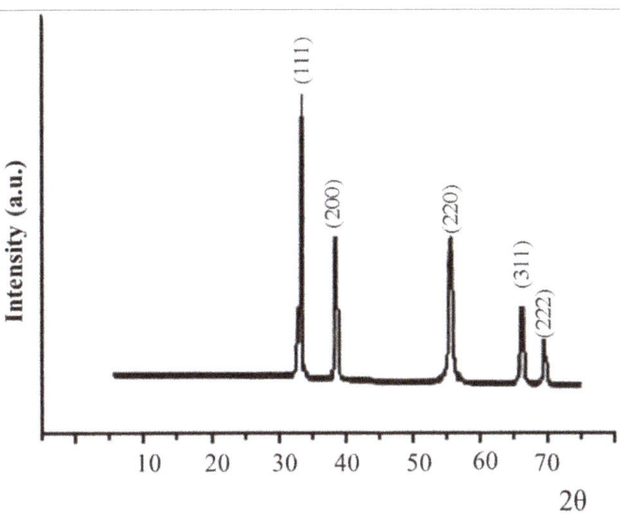

Figure 7. XRD pattern of cadmium oxide obtained by decomposition of (**1**).

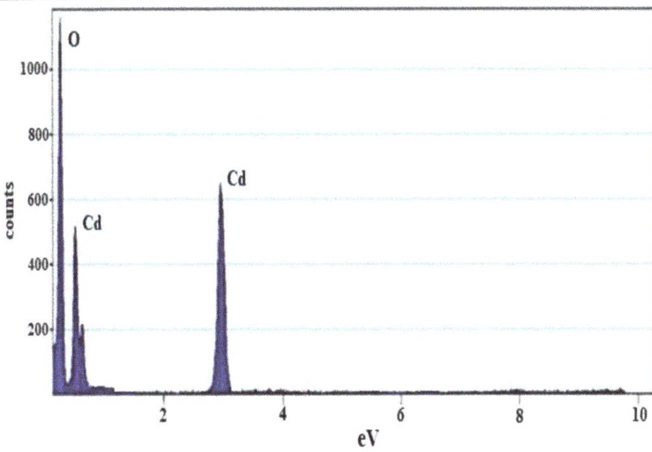

Figure 8. EDX of the as-prepared CdO nanoparticles.

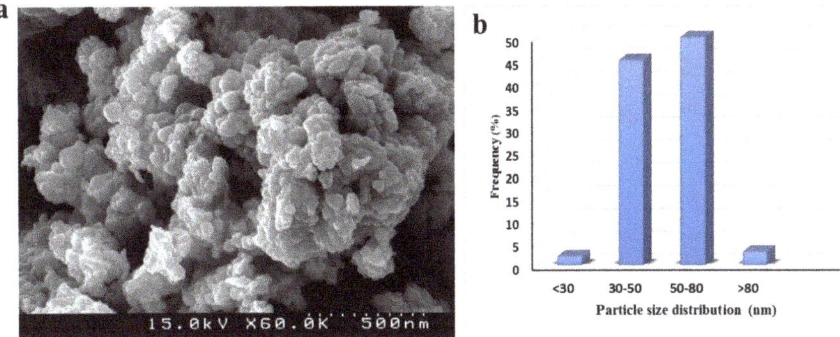

Figure 9. (**a**) SEM photograph and (**b**) size distribution graph of nano cadmium oxide (obtained by thermolysis of precursor **1**).

The XPS spectrum (Figure 10) confirms the presence of O and Cd in the sample. The binding energy corresponding to the peaks O1s, Cd 3d5/2, and 3d3/2 obtained by XPS analysis is 530.03, 405.6, and 412.9 eV, respectively [39].

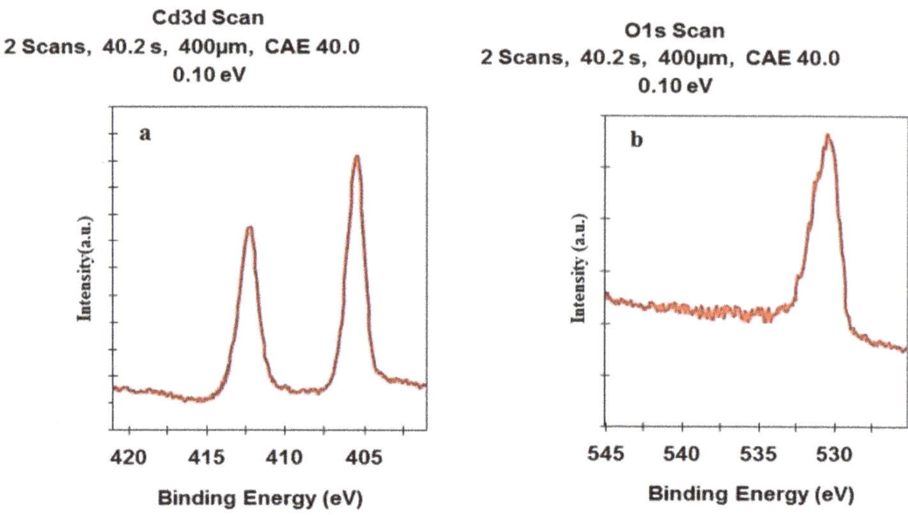

Figure 10. CdO XPS spectra (**a**) Cd scan and (**b**) O scan obtained by thermolysis of **1**.

The thermal gravimetric (TG) analysis was applied to evaluate the thermal stability of the $[Cd(L)(NO_2)_2]_n$ **1**. TG was recorded in the temperature range of 20 and 800 °C. The diagram of **1** depicts that the compound remains stable up to 125 °C, and then decomposes up to 210 °C. The first weight loss is related to the removal of the NO_2 unit. The step 2 in the range 300 °C to 380 °C with sharp weight loss is attributed to an organic moiety of **1** with a mass loss of about 45.3%. The remained solid around 800 °C is probably CdO (Figure 11).

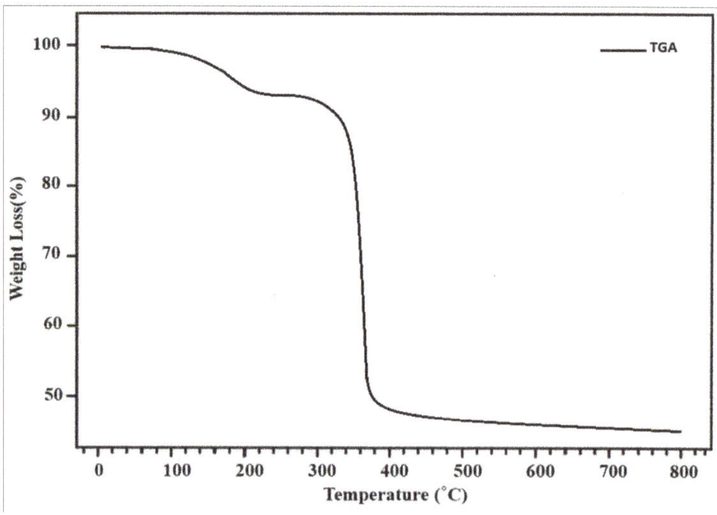

Figure 11. Thermal gravimetric (TG) thermogram of $[Cd(L)(NO_2)_2]_n$.

5. Conclusions

In the present study, a novel cadmium (II) coordination polymer $[Cd(L)(NO_2)_2]_n$ nanostructure was prepared in an ultrasound-assisted manner and also compared by its crystal structure. The structural characterization of new coordination compounds was carried out by elemental analysis, IR spectroscopy, SEM, and single-crystal crystallography. The complex's structure creates a 2D coordination polymer in solid-state. It was determined that the coordination number of Cd (II) ions is 7 with five oxygen of nitrite anions and two nitrogen atoms from two Schiff base ligand. With several weak interactions, the comprehensive system displayed a three-dimensional structure. Thermolysis of coordination polymer **1** results in CdO nanoparticles.

Supplementary Materials: The following are available online at https://www.mdpi.com/2073-4352/11/2/197/s1, Crystallographic data for the structure reported in this paper have been deposited with the Cambridge Crystallographic Data Centre as supplementary publication CCDC-2045446 for $[Cd(L)(NO_2)_2]_n$ (**1**). Copies of the data can be obtained by request from the CCDC, 12 Union Road, Cambridge CB2 1EZ, UK (Fax: +441223336033, eMail: deposit@ccdc.cam.ac.uk).

Author Contributions: Y.H.: supervision, writing—original draft; B.M.: investigation, writing—review and editing; J.D.: writing—review and editing. All authors have read and agreed to the published version of the manuscript.

Funding: This research was funded by grant number 32/1374 of the Sayyed Jamaleddin Asadabadi University.

Data Availability Statement: Not applicable.

Acknowledgments: The authors would like to express their gratitude to University of Qom for supporting this research.

Conflicts of Interest: The authors declare no conflict of interest.

References

1. Emhoff, K.A.; Balaraman, L.; Salem, A.M.H.; Mudarmah, K.I.; Boyd, W.C. Coordination chemistry of organic nitric oxide derivatives. *Coord. Chem. Rev.* **2019**, *396*, 124–140. [CrossRef]
2. Liu, X.; Hamon, J.R. Recent developments in penta-, hexa- and heptadentate Schiff base ligands and their metal complexes. *Coord. Chem. Rev.* **2019**, *389*, 94–118. [CrossRef]
3. Mirtamizdoust, B. Sonochemical synthesis of nano lead(II) metal-organic coordination polymer; New precursor for the preparation of nano-materials. *Ultrason Sonochem.* **2017**, *35*, 263–269. [CrossRef]
4. Rasheed, T.; Hassan, A.A.; Bilal, M.; Hussain, T.; Rizwan, K. Metal-organic frameworks based adsorbents: A review from removal perspective of various environmental contaminants from wastewater. *Chemosphere.* **2020**, *259*, 127369. [CrossRef]
5. Engel, E.R.; Scott, J.L. Advances in the green chemistry of coordination polymer materials. *Green Chem.* **2020**, *22*, 3693–3715. [CrossRef]
6. Mishra, A.; Jo, J.H.; Kim, H.; Woo, S.; Chi, K.W. A Discrete Cobalt Complex Obtained from a 1D Coordination Polymer for Highly Selective Sensing of the Mercury (II) Ion. *ChemPlusChem* **2014**, *79*, 925–928. [CrossRef]
7. Guo, Y.; Hu, X.; Zhang, X.; Pu, X.; Wang, Y. The synthesis of a Cu(II) Schiff base complex using a bidentate N_2O_2 donor ligand: Crystal structure, photophysical properties, and antibacterial activities. *RSC Adv.* **2019**, *9*, 41737. [CrossRef]
8. Chen, S.; Liu, X.; Ge, X.; Wang, Q.; Xie, Y.; Hao, Y.; Zhang, Y.; Zhang, L.; Shang, W.; Liu, Z. Lysosome-targeted iridium(III) compounds with pyridine-triphenylamine Schiff base ligands: Syntheses, antitumor applications and mechanisms. *Inorg. Chem. Front.* **2020**, *7*, 91–100. [CrossRef]
9. Che, C.M.; Huang, J.S. Metal complexes of chiral binaphthyl Schiff-base ligands and their application in stereoselective organic transformations. *Coord. Chem. Rev.* **2003**, *242*, 97–113. [CrossRef]
10. Himed, Y.; Komatsuzaki, N.O.; Sugihara, H.; Arakawa, H.; Kasuga, K. Transfer hydrogenation of a variety of ketones catalyzed by rhodium complexes in aqueous solution and their application to asymmetric reduction using chiral Schiff base ligands. *J. Mol. Catal A Chem.* **2003**, *195*, 95–100. [CrossRef]
11. Miroslaw, B. Homo-and Hetero-Oligonuclear Complexes of Platinum Group Metals (PGM) Coordinated by Imine Schiff Base Ligands. *Int. J. Mol. Sci.* **2020**, *21*, 3493. [CrossRef]
12. Vigato, P.A.; Tamburini, S. The challenge of cyclic and acyclic Schiff bases and related derivatives. *Coord. Chem. Rev.* **2004**, *248*, 1717–2128. [CrossRef]
13. Leovac, V.M.; Jevtovic, V.S.; Jovanovic, L.S.; Bogdanovic, G.A. Metal complexes with Schiff-base ligands—Pyridoxal and semicarbazide-based derivatives. *J. Serb. Chem. Soc.* **2005**, *70*, 393–422. [CrossRef]

14. Wilkinson, S.M.; Sheedy, T.M.; New, E.J. Synthesis and Characterization of Metal Complexes with Schiff Base Ligands. *J. Chem. Educ.* **2016**, *93*, 351−354. [CrossRef]
15. Masoomi, M.Y.; Morsali, A. Applications of metal–organic coordination polymers as precursors for preparation of nano-materials *Coord. Chem. Rev.* **2012**, *256*, 2921–2943. [CrossRef]
16. Malik, M.A.; O'Brien, P. Organometallic and Metallo-Organic Precursors for Nanoparticles. In *Precursor Chemistry of Advanced Materials*; Springer: Berlin/Heidelberg, Germany, 2005; p. 173.
17. Leong, W.L.; Vittal, J.J. One-Dimensional Coordination Polymers: Complexity and Diversity in Structures, Properties, and Applications. *Chem. Rev.* **2010**, *111*, 688–764. [CrossRef]
18. Vittal, J.J.; Ng, M.T. Chemistry of Metal Thio- and Selenocarboxylates: Precursors for Metal Sulfide/Selenide Materials, Thin Films, and Nanocrystals. *Acc. Chem. Res.* **2006**, *39*, 869–877. [CrossRef] [PubMed]
19. Das, K.; Datta, A.; Massera, C.; Sinha, C. Structural diversity, topology and luminescent properties of a two-dimensional Cd(II) coordination polymer incorporating 4,4′-dipyridyl and 4,4′-sulfonyldibenzoic. *J. Mol. Struct.* **2019**, *1179*, 618–622. [CrossRef]
20. Yadav, C.L.; Anamika; Rajput, G.; Kumar, K.; Drew, M.G.B.; Singh, N. Effect of Substituents on the Crystal Structures, Optical Properties, and Catalytic Activity of Homoleptic Zn(II) and Cd(II) β-oxodithioester Complexes. *Inorg. Chem.* **2020**, *59*, 11417–11431. [CrossRef] [PubMed]
21. Dutta, B.; Jana, R.; Sinha, C.; Ray, P.P.; Mir, M.H. Synthesis of Cd(II) based 1D coordination polymer by in situ ligand generation and fabrication of photosensitive electronic device. *Inorg. Chem. Front.* **2018**, *5*, 1998–2005. [CrossRef]
22. Mirtamizdoust, B.; Hanifehpour, Y.; Behzadfar, E.; Roodsari, M.S.; Jung, J.H.; Joo, S.W. A novel nano-structured three-dimensional supramolecular metal-organic framework for cadmium (II): A new precursor for producing nano cadmium oxide. *J. Mol. Struct.* **2020**, *1201*, 127191. [CrossRef]
23. Hanifehpour, Y.; Morsali, A.; Mirtamizdoust, B.; Joo, S.W. Sonochemical synthesis of tri-nuclear lead(II)-azido nano rods coordination polymer with 3,4,7,8-tetramethyl-1,10-phenanthroline (tmph): Crystal structure determination and preparation of nano lead(II) oxide. *J. Mol. Struct.* **2015**, *1079*, 67–73. [CrossRef]
24. Hanifehpour, Y.; Mirtamizdoust, B.; Hatami, M.; Khomami, B.; Joo, S.W. Synthesis and structural characterization of new bismuth (III) nano coordination polymer: A precursor to produce pure phase nano-sized bismuth (III) oxide. *J. Mol. Struct.* **2015**, *1091*, 43–48. [CrossRef]
25. Hanifehpour, Y.; Morsali, A.; Soltani, B.; Mirtamizdoust, B.; Joo, S.W. Ultrasound-assisted fabrication of a novel nickel(II)-bis-pyrazolyl borate twonuclear discrete nano-structured coordination compound. *Ultrason Sonochem.* **2017**, *34*, 519–524. [CrossRef] [PubMed]
26. Hammud, H.H.; Kortz, U.; Bhattacharya, S.; Demirdjian, S.; Hariri, E.; Isber, S.; Choi, E.S.; Mirtamizdoust, B.; Mroueh, M.; Daher, C.F. Structure, DFT studies, Magnetism and Biological activity of Bis[(μ 2-azido)-chloro-(1,10-phenanthroline)-copper(II)] complex. *Inorganica Chim. Acta.* **2020**, *506*, 119533. [CrossRef]
27. Kravtsov, V.C.; Lozovan, V.; Siminel, N.; Coropceanu, E.B.; Kulikova, O.V.; Costriucova, N.V.; Fonari, M.S. Fonari, From 1D to 2D Cd(II) and Zn(II) coordination networks by replacing monocarboxylate with dicarboxylates in partnership with azine ligands: Synthesis, crystal structures, inclusion, and emission properties. *Polyhedron* **2020**, *180*, 114411. [CrossRef]
28. Mercury 2.4, Copyright Cambridge Crystallographic Data Centre: 12 Union Road. Cambridge, UK; pp. 2001–2010.
29. Oxford Diffraction. *CrysAlis RED and CrysAlis CCD software (Ver. 1.171.32.5)*; Oxford Diffraction Ltd.: Abingdon, Oxfordshire, UK, 2010.
30. Pal, A.; Chand, S.; Elahi, S.M.; Das, M.C. A microporous MOF with a polar pore surface exhibiting excellent selective adsorption of CO_2 from CO_2–N_2 and CO_2–CH_4 gas mixtures with high CO_2 loading. *Dalton Trans.* **2017**, *46*, 15280–15286. [CrossRef]
31. Dong, Y.B.; Smith, M.D.; Loye, H.C.Z. New Inorganic/Organic Coordination Polymers Generated from Bidentate Schiff-Base Ligands. *Inorg. Chem.* **2000**, *39*, 4927–4935. [CrossRef]
32. Harrowfield, J.M.; Miyamae, H.; Skelton, B.W.; Soudi, A.A.; White, A.H. Lewis-Base Adducts of Lead (II) Compounds. XX. Synthesis and Structure of the 1:1 Adduct of Pyridine With Lead (II) Thiocyanate. *Aust. J. Chem.* **1996**, *49*, 1165–1169. [CrossRef]
33. Patterson, A.L. The Scherrer formula for X-ray particle size determination. *Phys. Rev.* **1939**, *56*, 978. [CrossRef]
34. Gao, E.Q.; Cheng, A.L.; Xu, Y.X.; Yan, C.H.; He, M.Y. New Inorganic–Organic Hybrid Supramolecular Architectures Generated from 2, 5-Bis (3-pyridyl)-3, 4-diaza-2, 4-hexadiene. *Cryst. Growth Des.* **2005**, *5*, 1005–1011. [CrossRef]
35. Lee, G.H.; Wang, H.T. Synthesis and Crystal Structure of a Three-Dimensional Nickel (II) Coordination Polymer with 1, 4-Bis (3-pyridyl)-2, 3-diazo-1, 3-butadiene as a Ligand. *Anal. Sci. X-ray Struct. Anal. Online* **2007**, *23*, x1–x2. [CrossRef]
36. Soltanzadeh, N.; Morsali, A. Metal–organic supramolecular assemblies generated from bismuth (III) bromide and polyimine ligands. *Polyhedron* **2009**, *28*, 703–710. [CrossRef]
37. Khanpour, M.; Morsali, A. Solid state crystal-to-crystal transformation from a monomeric structure to 1-D coordination polymers on anion exchange. *CrystEngComm* **2009**, *11*, 2585–2587. [CrossRef]
38. Mahmoudi, G.; Morsali, A. Three new HgII metal–organic polymers generated from 1, 4-bis (n-pyridyl)-3, 4-diaza-2, 4-hexadiene ligands. *Inorg. Chim. Acta* **2009**, *362*, 3238–3246. [CrossRef]
39. Velusamy, P.; Ramesh Babu, R.; Ramamurthi, K.; Elangovan, E.; Viegas, J.; Dahlem, M.S.; Arivanandhan, M. Characterization of spray pyrolytically deposited high mobility praseodymium doped CdO thin films. *Ceram. Int.* **2016**, *42*, 12675–12685. [CrossRef]

Article

Green Synthesis, SC-XRD, Non-Covalent Interactive Potential and Electronic Communication *via* DFT Exploration of Pyridine-Based Hydrazone

Akbar Ali [1,2,3,*,†], Muhammad Khalid [4,†], Saba Abid [4], Muhammad Nawaz Tahir [5], Javed Iqbal [6], Muhammad Ashfaq [5], Fariha Kanwal [7], Changrui Lu [1,*] and Muhammad Fayyaz ur Rehman [3]

1 Department of Chemistry, Chemical Engineering and Biotechnology, Donghua University, Shanghai 201620, China
2 Department of Chemistry, University of Malakand, Chakdara, Lower Dir, Khyber Pakhtunkhwa 18800, Pakistan
3 Department of Chemistry, University of Sargodha, Sargodha 40100, Pakistan; fayyaz9@gmail.com
4 Department of Chemistry, Khwaja Fareed University of Engineering & Information Technology, Rahim Yar Khan 64200, Pakistan; muhammad.khalid@kfueit.edu.pk (M.K.); sabaabid206@gmail.com (S.A.)
5 Department of Physics, University of Sargodha, Sargodha, Punjab 40100, Pakistan; dmntahir_uos@yahoo.com (M.N.T.); ashfaq.muhammad@uos.edu.pk (M.A.)
6 Department of Chemistry, University of Agriculture, Faisalabad 38000, Pakistan; javed.iqbal@uaf.edu.pk
7 Med-X Research Institute, School of Biomedical Engineering, Shanghai Jiao Tong University, Shanghai 200240, China; Farihakaanwal@gmail.com
* Correspondence: akbarchm@gmail.com (A.A.); crlu@dhu.edu.cn (C.L.)
† Both authors contributed equally to this work.

Received: 19 July 2020; Accepted: 28 August 2020; Published: 2 September 2020 

Abstract: Ultrasound-based synthesis at room temperature produces valuable compounds greener and safer than most other methods. This study presents the sonochemical fabrication and characterization of a pyridine-based halogenated hydrazone, (E)-2-((6-chloropyridin-2-yl)oxy)-N′-(2-hydroxybenzylidene) acetohydrazide (HBPAH). The NMR spectroscopic technique was used to determine the structure, while SC-XRD confirmed its crystalline nature. Our structural studies revealed that strong, inter-molecular attractive forces stabilize this crystalline organic compound. Moreover, the compound was optimized at the B3LYP/6-311G(d,p) level using the Crystallographic Information File (CIF). Natural bonding orbital (NBO) and natural population analysis (NPA) were performed at the same level using optimized geometry. Time-dependent density functional theory (DFT) was performed at the B3LYP/6-311G (d,p) method to calculate the frontier molecular orbitals (FMOs) and molecular electrostatic potential (MEP). The global reactivity descriptors were determined using HOMO and LUMO energy gaps. Theoretical calculations based on the Quantum Theory of Atoms in Molecules (QT-AIM) and Hirshfeld analyses identified the non-covalent and covalent interactions of the HBPAH compound. Consequently, QT-AIM and Hirshfeld analyses agree with experimental results.

Keywords: hydrazones; sonochemical-based synthesis; single-crystal analysis; non-covalent interaction; Hirshfeld surface study

---

1. Introduction

Humanity faces increasing health, shelter, and economic problems as we consume more resources to pollute, urbanize, and deforest our environment. Fatal diseases have not only taken many lives but also severely harmed the global economy. To combat global pandemics and other diseases, synthetic organic chemists need to synthesize novel and potent chemicals by safe and green methods. One such chemical, hydrazones, plays a substantial role in the bio-medicinal applications due to its versatility [1–3]. It has many effortlessly reachable binding sites for the medicinal applications [4], such as antimicrobial [5], cardioprotective [6], anti-HIV [7], anti-inflammatory [8], anticancer [9], antihypertensive [10], antitubercular [11], antimalarial [12], antidepressant [13], antioxidant [14], and anticonvulsant [15]. For example, pyridine-based hydrazone derivatives ubiquitously displayed antifungal properties [16–18]. Acylpyridine derivative, 2-benzoxazolylhydrazon, suppresses leukemia, colon and ovarian cancer cell lines [19]. Acylhydrazone introduction in 1, 2, 4-triazolo [4, 3-a] pyridine derivatives by a microwave-assisted method leads to herbicidal and pesticidal lead compounds [1]. Hydrazone derivatives also possess unique physical and chemical properties including fluorescence emission [20], corrosion inhibitory properties and passivation [21], and iron chelation in iron toxicity [22]. In addition, lone electron pairs and pi-electrons play a key role in medicinal applications due to their ability of non-covalent interaction, such as van der Waals interactions, hydrophobic bonds, ionic bonds, and hydrogen bonds [23–26]. In particular, these non-covalent interactions facilitate crystals packing, proton transfer reactions, the stability of molecules, enzymatic catalysis [25,27–29]. Several molecules with hydrogen bonding capacity are important for catalysis in organic transformation such as diols, bisphenols, hydroxy acids, urea, guanidinium and amidinium ions, thioureas, lactams, thioureas, cinchona alkaloids, and phosphoric acids [30–34]. Amongst them, hydrazones have a unique chemical architecture (Figure 1), allowing its significant ability to form non-covalent interactions [35].

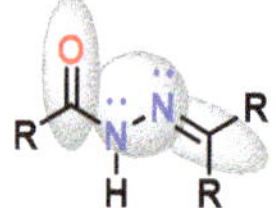

Figure 1. The functional diversity of the hydrazone skeleton.

Studies show that microwave-assisted synthesis accelerates chemical synthesis with a better yield and higher purity in comparison to conventional methods [18,36,37]. Microwave (MW) radiations assist in non-thermal polarizing radiation, dipolar polarization, ionic conduction reactions [38]., This study reports the ultrasound-based synthesis, SC-XRD exploration, and density functional theory (DFT) analysis of the pyridine-based novel crystalline hydrazones, i.e., (E)-2-((6-chloropyridin-2-yl)oxy)-N′-(2-hydroxybenzylidene) acetohydrazide.

2. Materials and Methods

2.1. General

Analytical grade solvents and pure reagents were used without any further purification. TLC (Thin layer chromatography) cards, coated with silica gel (0.25 mm thickness), were used to monitor the reaction progress. For the NMR spectra measurement, Bruker-Avance, A-V spectrometer, was used. For the single crystal analysis, Bruker Kappa APEX-II diffractometer was used where the data correction and data reduction were made by APEX-II and SAINT, respectively [39]. For the structure solution, SHELXS97 software [40,41] and for refinement, SHELXL2014/6 was used to minimize the structural errors [42]. For the graphical representation of the asymmetric unit, ORTEP was used while for the hydrogen bonding, PLATON was used [43].

2.2. *Synthesis of 2-(6′-Chloroazin-2′-yl) oxy-aceto-hydrazide (A)*

The precursor A was manufactured according to the procedure described elsewhere [35,44]. Accordingly, a mixture of ethyl 2-(6′-chloroazin-2-yl)-ox-ethanoate (131 mg, 0.61 mmol) and $N_2H_4.H_2O$ (0.09 mL, 1.83 mmol) in ethanol was refluxed for 3 h. The reaction on completion (monitored by TLC) was cooled to room temperature and concentrated under reduced pressure. The targeted hydrazide was purified by column chromatography yielding 89 mg of the isolated A (73%).

2.3. General Procedure for the Synthesis of (E)-2-((6-Chloropyridin-2-yl)oxy)-N′-(2-hydroxybenzylidene)acetohydrazide

A mixture of A (2-(6′-chloroazin-2′-yl) oxy-aceto-hydrazide) (0.48 mmol) and Salicylaldehyde (0.54 mmol) was dissolved in ethanol separately to make clear solutions. The solutions were mixed at room temperature, and the mixture was sonicated for 5 to 10 min. The targeted compound was precipitated that was filtered through standard filtration and recrystallized in ethanol (Scheme 1).

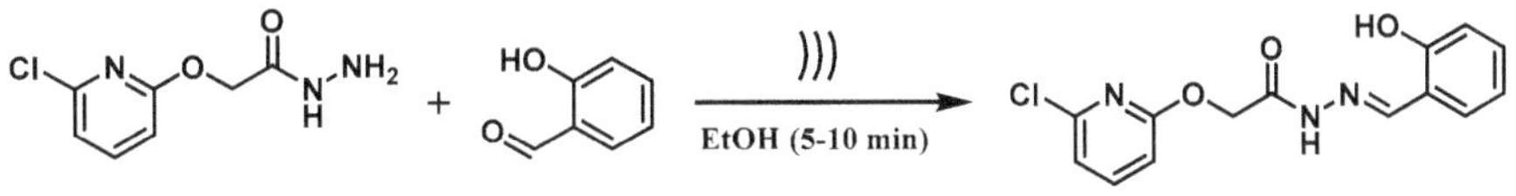

Cl N O O N NH$_2$ H + HO O))) EtOH (5-10 min) Cl N O O N N H HO

A 85%

Scheme 1. Ultrasonic-based Synthesis of (E)-2-((6-chloropyridin-2-yl)oxy)-N′-(2-hydroxybenzylidene) acetohydrazide.

^{1}H NMR (400 MHz, DMSO) δ 11.84 (s, 1H), 10.99 (s, 1H), 8.48 (s, 1H), 7.81 (dt, J = 15.3, 7.8 Hz, 2H), 7.72–7.67 (m, 1H), 7.32–7.21 (m, 2H), 7.15 (d, J = 7.5 Hz, 1H), 7.00–6.94 (m, 1H), 5.31 (s, 2H). ^{13}C NMR (100 MHz, DMSO) δ 168.2, 163.8, 162.4, 156.4, 147.6, 142.1, 141.3, 131.2, 126.2, 119.3, 116.8, 116.1, 109.6, 62.9 (Figure S1).

2.4. Computational Studies

The simulation study for the entitled compound, (E)-2-((6-chloropyridin-2-yl)oxy)-N′-(2-hydroxybenzylidene) acetohydrazide (HBPAH), was performed through DFT [45–47] employing Gaussian 09 program package [48]. By the use of GaussView 5.0. [49] all input files were organized. Finally, Chem craft [50], Avogadro [51] and Gauss Sum [52], AIM-All Professional [53], and Crystal Explorer [54] programs were used for the interpretation of output files. The Structure of HBPAH was optimized using SC-XRD-based geometry at the B3LYP/6-311G(d,p) level. The natural bond orbital (NBO) [55,56] and natural population analysis (NPA) were analyzed similarly, while the frontier molecular orbital (FMO) and molecular electrostatic potential (MEP) were calculated by the TD-DFT/B3LYP /6-311G(d,p) level [57,58]. Moreover, the Hirshfeld surface (HS) analysis [59,60] was carried out to determine the non-covalent interactions. The Quantum Theory of Atoms in Molecules (QT-AIM) [61] analysis was employed to explore the non-covalent interactions. The electron affinity (*EA*), electronegativity (*X*) [62], global electrophilicity index (ω) [62–64], ionization potential (*IP*) [65], global hardness (η) [66,67], global softness (*S*) [68] and chemical potential (μ) [69] were known as global reactivity parameters, and their values can be calculated by HOMO-LUMO energies. These parameters were also reported as biological activity descriptors having numerous optoelectronic applications and are helpful in determining stability, reactivity, and selectivity of the molecules [62,70–72]. They were calculated through Equations (1)–(7):

$$IP = -E_{HOMO} \tag{1}$$

$$EA = -E_{LUMO} \tag{2}$$

where *IP* = ionization potential (eV), *EA* = electron affinity (eV).

Koopmans's theorem [73] was usually used to calculate the chemical potential (μ), electronegativity (x) and chemical hardness (η) and was equated as:

$$\mu = \frac{E_{HOMO} + E_{LUMO}}{2} \tag{3}$$

$$x = \frac{[IP + EA]}{2} = -\frac{[E_{LUMO} + E_{HOMO}]}{2} \tag{4}$$

$$\eta = \frac{[IP - EA]}{2} = -\frac{[E_{LUMO} - E_{HOMO}]}{2} \tag{5}$$

The following equation was used for global softness (σ):

$$\sigma = \frac{1}{2\eta} \tag{6}$$

The calculation of electrophilicity index (ω) was reported by Parr et al. as:

$$\omega = \frac{\mu^2}{2\eta} \tag{7}$$

3. Results and Discussion

The hydrazone, (E)-2-((6-chloropyridin-2-yl)oxy)-N′-(2-hydroxybenzylidene) acetohydrazide (HBPAH), was synthesized with a yield of 85% and its structures were determined by NMR spectroscopy and SC-XRD analysis. The ^{1}H- and ^{13}C-NMR of the title compound showed the presence of each signal in duplication that indicates that the title compound exists in two isomeric forms; a minor isomer A (E) that is 45.87% and a major isomer B (Z) that is 54.12% (Scheme 2). The ratio of the E and Z isomers was calculated from the ^{1}H-NMR analysis, where the methylenic signals of both isomers were integrated into the ^{1}H NMR spectra (Figure S2 in Supplementary Materials).

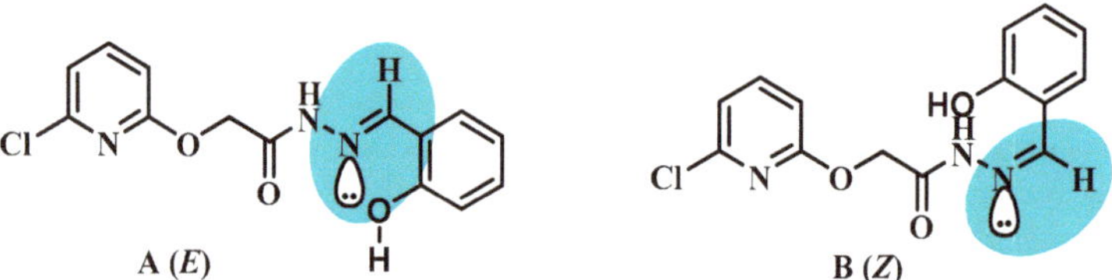

Scheme 2. Isomeric existence of the title compound. The highlighted region shows the double-bond conferring E and Z isomers.

The DFT calculation of HBPAH was performed by DFT/B3LYP/6-311G (d, p). Table 1 shows the single-crystal analysis details, Hirshfeld surface, and computational details.

Table 1. Experimental details of the compound (E)-2-((6-chloropyridin-2-yl)oxy)-N′-(2-hydroxybenzylidene) acetohydrazide (HBPAH).

Crystal Data	HBPAH
CCDC* number	2012169
Chemical formula	$2(C_{14}H_{12}ClN_3O_3)\cdot H_2O$
M_r	629.45
Crystal system, space group	Triclinic, $P\bar{1}$
Temperature (K)	296
a, b, c (Å)	6.6987 (8), 7.3628 (9), 31.513 (4)
α, β, γ (°)	90.978 (7), 93.508 (6), 113.954 (7)
V (Å^3)	1416.2 (3)
Z	2
Density (calculated)	1.476 Mg/m^3
F(000)	652

Table 1. *Cont.*

Crystal Data	HBPAH
Radiation type	Mo $K\alpha$
Wavelength (λ)	0.71073 Å
μ (mm^{-1})	0.288
Crystal shape	Needle
Crystal Color	Colorless
Crystal size (mm)	0.38 × 0.22 × 0.18
Data Collection	HBPAH
Diffractometer	Bruker APEXII CCD diffractometer
Absorption correction	multi-scan (SADABS; Bruker, 2007)
No. of measured, independent and observed [$I > 2s(I)$] reflections	15,051, 5452, 2920
R_{int}	0.070
Theta range for data collection	0.648 to 26.000°
Index ranges	$-8 \leq h \leq 7, -9 \leq k \leq 9, -37 \leq l \leq 38$
$(\sin \theta/\lambda)_{max}$ ($Å^{-1}$)	0.617
Refinement	HBPAH
$R[F^2 > 2\sigma(F^2)], wR(F^2), S$	0.083, 0.195, 1.05
No. of reflections	5452
No. of parameters	396
H-atom treatment	H atoms treated by a mixture of independent and constrained refinement
$\Delta\rho_{max}, \Delta\rho_{min}$ (e $Å^{-3}$)	0.24, −0.30

*CCDC (Cambridge Crystallographic Data Centre).

HBPAH (Table 1, Figure 2) crystals contain two crystallographically independent molecules of (E)-2-((6-chloropyridin-2-yl)oxy)-N′-(2-hydroxybenzylidene)acetohydrazide and one water molecule. In the first molecule (C1-C14/N1-N3/O1-O3/CL1) (red in overlay plot), the 6-chloropyridin-2-ol moiety A (C1-C5/N1/CL1), acetohydrazide group B (C6/C7/N2/N3/O2) and O-cresol moiety C (C8-C14/O3) are planar with an r.m.s deviation of 0.0058, 0.0067 and 0.0133 Å, respectively, whereas, in the second molecule (C15-C28/N4-N6/O4-O6/CL2) (blue in overlay plot), the similar moieties D (C15-C19/N4/O4/CL2), E (C20/C21/N5/N6/O5) an F (C22-C28/O6) are planar with r.m.s deviation of 0.0065, 0.0034, and 0.0108 Å, respectively. The dihedral angles between moieties in the first molecule A/B, A/C, and B/C are 14.15 (1)°, 9.82 (1)°, and 23.13 (1)°, respectively whereas the dihedral angle between similar moieties in second molecule D/E, D/F, and E/F is 12.37 (1)°, 11.87 (1)° and 3.7 (1)°, respectively. The two crystallographic independent molecules differ in terms of geometric parameters, as shown in Figure 3. The second molecule is inverted and then made to overlap with the first molecule. This analysis shows that the root mean square deviation between the first molecule and the second molecule is 0.2376 Å.

In both molecules within the lattice, the NH of acetohydrazide group interacts with the O-atom of 6-chloropyridin-2-ol moiety through intra N-H···O bonding to form S(5) loop, and the hydroxyl group of o-cresol moiety interacts with N-atom of acetohydrazide group through intra O-H···N bonding to form S(6) loop. The first molecule connects with the second molecule through N-H···O bonding, where NH is from acetohydrazide group E, and O-atom is from the acetohydrazide group B. Water molecule is engaged in two types of classical H-bonding named as O-H···O and N-H···O. Water acts as a donor in O-H···O (carbonyl O-atom of acetohydrazide group B) and O-H···O (carbonyl O-atom of acetohydrazide group E) to connect molecule of the first type with a molecule of the second type whereas it acts as an acceptor in N-H···O bonding where NH is from acetohydrazide group E. Water molecule is also engaged in one weak non-classical C-H···O (CH is from O-cresol moiety C) bonding with C-O distance of 3.271 Å and angle of 139.06° [74,75]. $R_2^1(6)$ loop is formed through classical N-H···O and non-classical C-H···O bonding in which water acts as an acceptor. The carbonyl O-atom of acetohydrazide group E is also engaged in weak non-classical C-H··· O(CH is from o-cresol moiety F) bonding to connect molecules of the second type with each other with a C-O distance of 3.433 Å and angle of 162.40° [76].

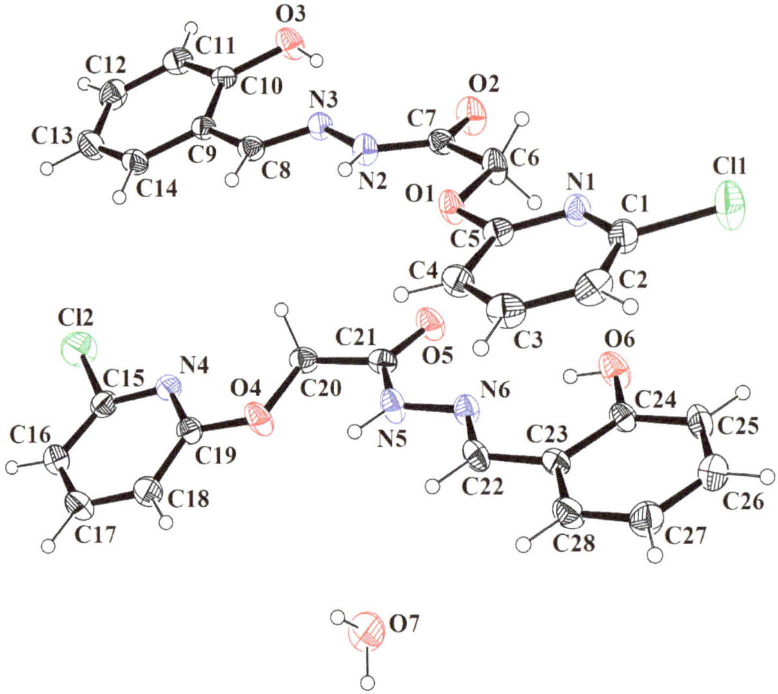

Figure 2. ORTEP diagram of HBPAH drawn at a probability level of 30% with H-atoms are displayed by tiny circles of arbitrary radii. Red color shows oxygens, blue nitrogen, green chlorine, white is for hydrogen, and black/white contours show carbon atoms.

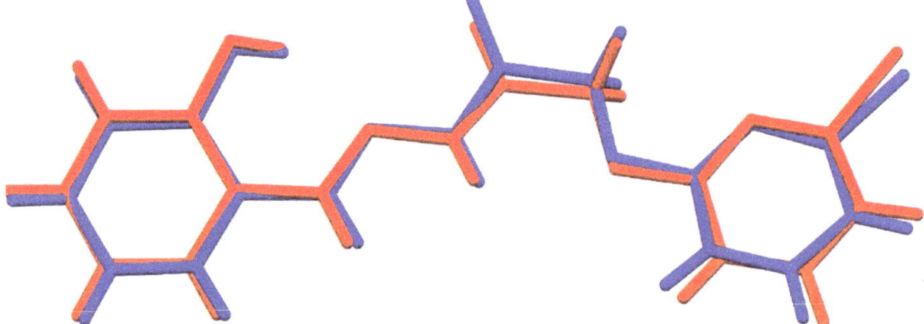

Figure 3. Molecular overlay of two crystallographically independent molecules: first molecule (red) and second molecule (blue).

All the above-mentioned loops and H-bonding are shown in Figure 4, Figure S3 and Table 2. Both molecules and water are connected to form an infinite 2-D network in the crystallographic plane (0 0 1) with base vector (1 0 0) and (0 1 0). Along with the intra and intermolecular H-bonding, a cyclic face-to-face stacking between different rings assists in further strengthening crystal packing.

The pyridine ring (C1-C5/N1) at the asymmetric position stacks with two symmetry mates' phenyl rings (C23-C28) located at (x, 1 + y, z) and (1 + x, 1 + y, 1 + z) with inter-centroid separation of 3.671 Å and 3.810 Å as displayed in Figure 5 and Table 3. Similarly, the phenyl ring (C9-C14) at the asymmetric position stacks with two symmetry-related pyridine rings (C15-C19/N4) located at (1 + x, 1 + y, 1 + z) and (1 + x, 1 + y, z) with inter-centroid separation of 3.874 Å and 3.876 Å, respectively. Cg(1), Cg(2), Cg(3), and Cg(4) are the centroids of pyridine ring (C1-C5/N1), phenyl ring (C9-C14), pyridine ring (C15-C19/N4), phenyl ring (C23-C28), respectively. Dde, DAde, De (f) and Df (e), respectively, show the distance between centroids of rings, the dihedral angle between the planes of rings, perpendicular distance of Cg(e) to Cg(f), perpendicular distance of Cg(f) to Cg(e).

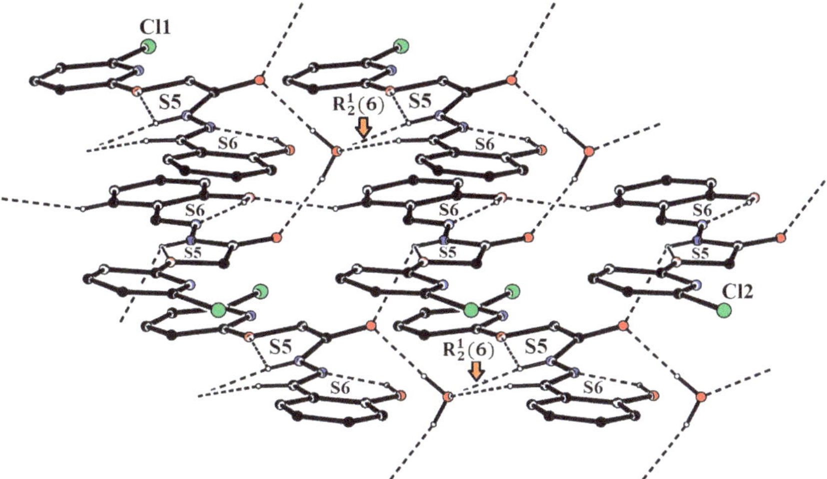

Figure 4. Packing diagram of HBPAH showing H-bonded connection of first, 2nd type of molecules and water with H-atoms not engaged in H-bonding are omitted for clearness. Red color shows oxygens, blue nitrogen, green chlorine, and black/white contours shows carbon atoms.

Table 2. Geometrical parameters of potential Hydrogen-bonds (Å, °) for HBPAH.

D–H···A	D–H	H···A	D···A	D–H···A
O3–H3*A*···N3	0.82	1.95	2.659 (5)	145
N2–H2*A*···O1	0.86	2.235	2.655	105.34
N2–H2*A*···O7	0.86	2.08	2.902 (5)	160
O6–H6···N6	0.82	1.88	2.587 (5)	145
N5–H5···O4	0.86	2.170	2.571	108.18
N5–H5···O2^{i}	0.86	2.48	3.029 (5)	122
O7–H7*A*···O5^{ii}	0.91 (6)	2.03 (6)	2.877 (5)	155 (6)
O7–H7*B*···O2^{ii}	0.89 (6)	1.95 (7)	2.826 (6)	168 (6)

Symmetry codes: ^{i} x − 1, y − 1, z; ^{ii} x − 1, y, z.

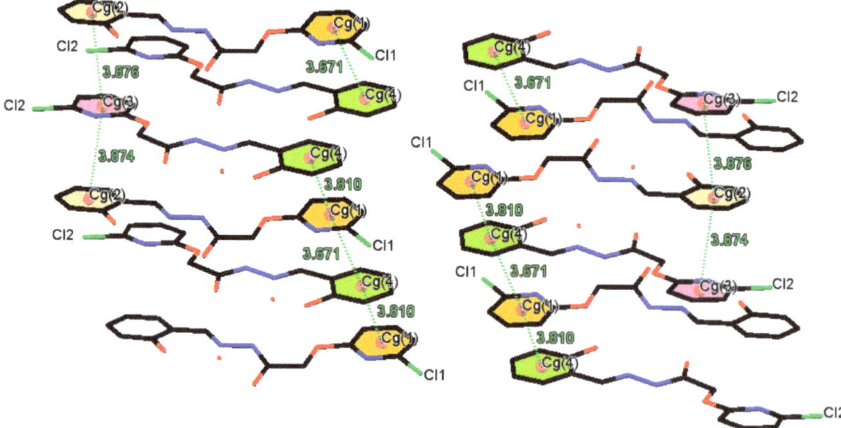

Figure 5. Cyclic Face-to-face stacking interaction between various rings in the crystal packing. Distances shown are in Å with H-atoms omitted for clarity.

Table 3. Geometry-related parameters of cyclic face-to-face stacking interactions for HBPAH with distance given in Å.

Cg(e)–Cg(f)	D_{ef}	DA_{ef}	D_e (f)	D_f (e)	Ring Off-Set
Cg(1)–Cg(4)^{iii}	3.671	1.2(2)	3.4556(19)	3.460(2)	-
Cg(1)–Cg(4)^{iv}	3.810	1.2(2)	3.4495(19)	3.479(2)	-
Cg(2)–Cg(3)^{iv}	3.874	2.7(2)	3.4350(19)	3.3581(19)	-
Cg(2)—Cg(3)^{v}	3.876	2.7(2)	3.4495(19)	3.4066(19)	-

Symmetry codes: ^{iii} x, 1 + y, z; ^{iv} 1 + x, 1 + y, 1 + z; ^{v} 1 + x, 1 + y, z.

3.1. Comparative Structural Study

The SC-XRD-based structure of HBPAH was used for geometry optimization in bond length and bond angle calculations. For HBPAH, an atom numbering scheme was presented in Figure S4 (Supplementary Information), and the aforementioned geometrical parameter results were shown in Table S1 (Supplementary Information). DFT-calculated and SC-XRD-driven parameters agree with each other with an overall variation of 0.039 ± 0.028 Å. Similarly, bond angles in HBPAH deviate around 3.0 ± 3.3°.

3.1.1. Hirshfeld Surface Analysis

The crystal structure of HBPAH contains many N-H/O, C-H/O, O-H/O, C-O$\cdots\pi$, C-H$\cdots\pi$, and $\pi\cdots\pi$ interactions. The HS analysis calculates the percentage of significant non-covalent interactions contributions [77–82]. The HS mapped with properties like d_{norm}, d_e, d_i, shape index, curvedness, and the 2D fingerprint plots of HBPAH are shown in Figure 5, Figures S5 and S6 (Supplementary Information). Red and white in the HS analysis represent the strongest and intermediate interactions, whereas blue illustrates weaker intermolecular interactions. As d_e and d_i are external and internal distance from a surface to the nearest nuclei, respectively, d_{norm} can be defined by Equation (8) [83]:

$$d_{norm} = \frac{d_i - r_i^{vdw}}{r_i^{vdw}} + \frac{d_e + r_e^{vdw}}{r_e^{vdw}} \tag{8}$$

In HBPAH, the d_{norm} surfaces with dark red spots demonstrate hydrogen bonding interactions [84,85]. The oxygen of –C=O aceto group, the nitrogen of hydrazide -NH, the hydroxyl group of *N′*-(2-hydroxybenzildene), and other hydrazide nitrogen near the *N′*-(2-hydroxybenzildene) participate the strong interactions as shown in Figure 6. The HS analysis of HBPAH has mapped the distances d_{norm} (−0.5807 to 1.0525 *a.u.*), shape index (−1.000 to 1.000 *a.u.*), and curvedness (−4.000 to 0.4000 *a.u.*), as shown in Figure 6.

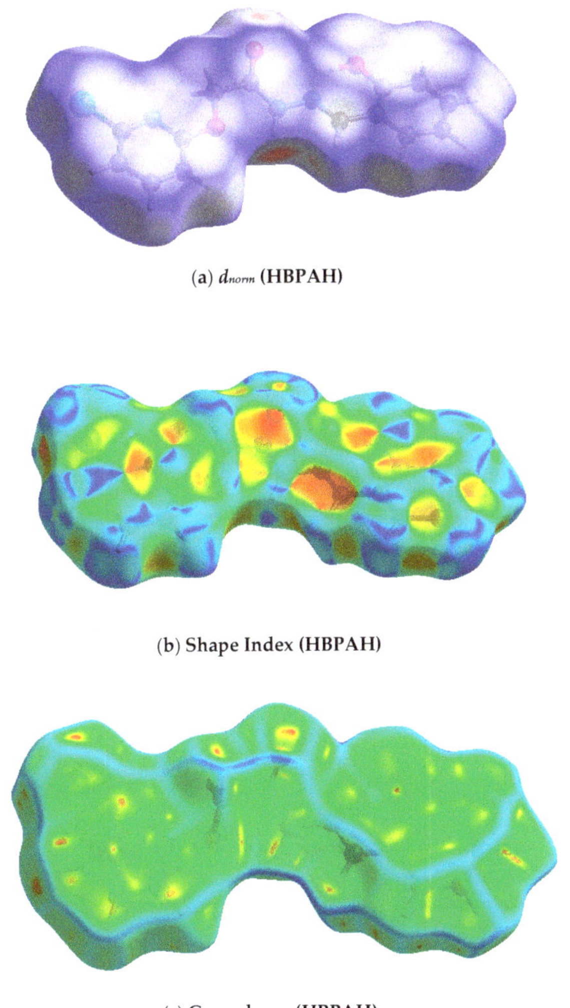

(**a**) d_{norm} **(HBPAH)**

(**b**) **Shape Index (HBPAH)**

(**c**) **Curvedness (HBPAH)**

Figure 6. Hirshfeld surfaces of the entitled compound mapped over (**a**) d_{norm}, (**b**) shape index, and (**c**) curvedness, respectively, for HBPAH (1 a.u. of electron density = 6.748 e.Å^{-3}).

In the curvedness diagram, the broader green areas separated by blue outlines show the stacking interactions. Figure 6 shows the shape index that explains the π–π stacking interactions with blue humps and red hollows.

We then used two-dimensional fingerprint plots to explain the intermolecular interactions within the molecular structure [86–88]. The strongest interaction among hydrogen atoms in the compound is 33.2%, as shown in Figure 7, alongside percentage contribution for all interatomic contacts. Figure S5 shows the two-dimensional fingerprint plots. The most dominant contributions within the crystal packing are as follows: H-H (33.20%), C-H (13.00%), O-H (17.20%), Cl-H (15.60%), C-C (7.50%) and C-N

(2.70%). Our HS analysis shows that C-H$\cdots\pi$ interactions dominate the stability within the molecular structure of HBPAH.

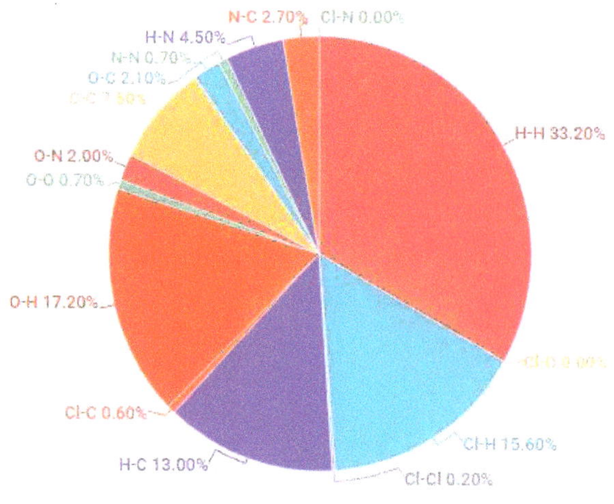

HBPAH.

Figure 7. Percentage contributions of all interatomic contacts for an entitled compound.

Our HS analysis also reports secondary interactions between molecules [78,87,89], such as carbon atom attached with –NH of hydrazide part bonded with the hydrogen atom of the O=C-H group [90]. Figure 8 shows the intermolecular hydrogen bonds (dashed green lines between the hydrazide –NH and the hydroxybenzylide O-H) and intermolecular hydrogen bond with the water molecule (solvent interaction).

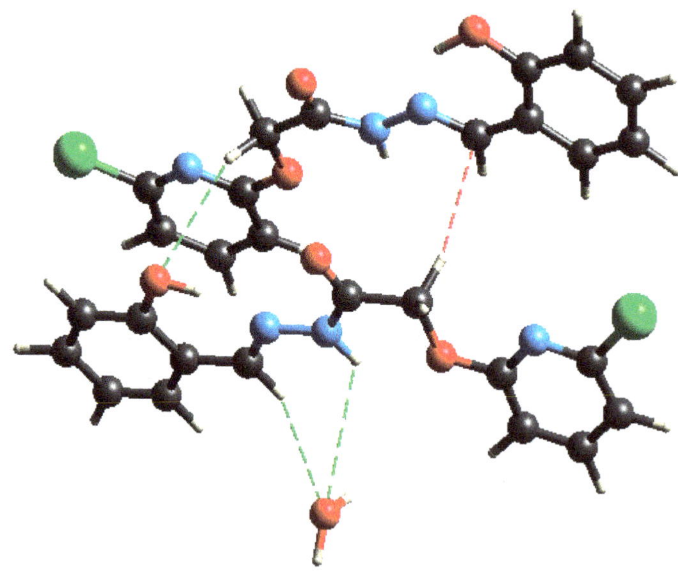

Figure 8. Hydrogen bonds in HBPAH.

3.1.2. QT-AIM Analysis

Next, we used the Quantum Theory of Atoms in Molecules (QT-AIM) [61,91,92] to analyze non-covalent inter and intramolecular interactions, such as hydrogen bonding (HBs) (Table S2, Supplementary Information). The AIM analysis revealed that the crystal is stabilized through intra- and inter-molecular interactions [93,94], as shown by the dashed bond paths (BPs) (Figure 9). We calculated the non-covalent interactions (NCI) by calculating real-space regions where non-covalent interactions are essential and based entirely on ρ and its gradient [94]. HBPAH displayed intermolecular interactions that stabilized the molecules within the crystal. The O-H ρ values at BCPs (Bond critical points), H16-O36, H8-O36, H23-O36 and H19-O37 were +0.0029 e/a^3, +0.0127 e/a^3, +0.0091 e/a^3 and +0.0073 e/a^3, respectively. The N-H ρ values at BCPs, H5-N9, and H38-N42 were +0.0436 e/a^3 and +0.0421 e/a^3, respectively (Table 4). Other intermolecular interactions, O2-O37 and H33-H53 were +0.0066 e/a^3 and +0.0016 e/a^3.

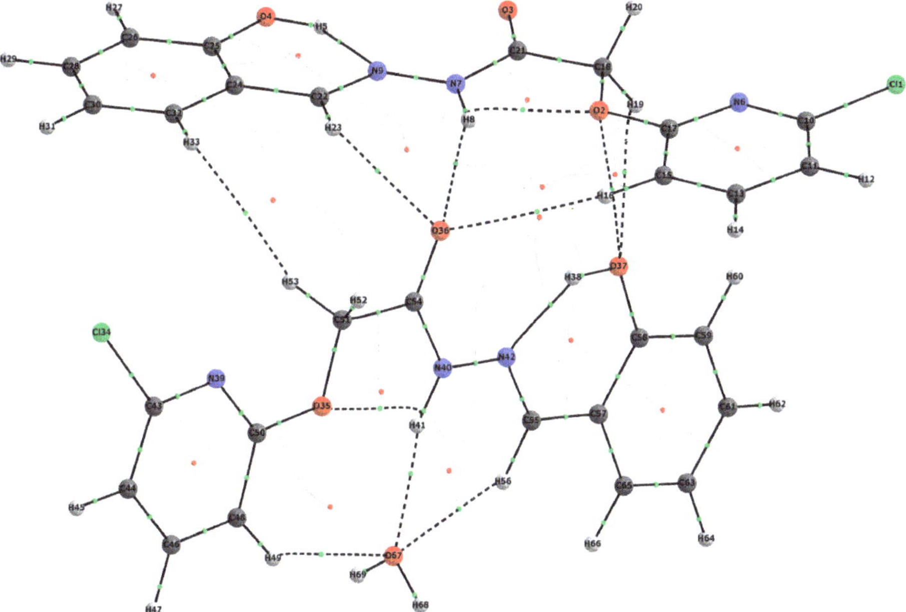

Figure 9. AIM based Schematic structure of HBPAH.

Table 4. AIM properties of selected values for the main interactions for HBPAH; electronic density (ρ), Laplacian of density ($\nabla^2\rho$), ellipticity (ε) and density of potential energy (V), Laplacian of density ($\nabla^2\rho$), ellipticity (ε) and density of potential energy (V).

BCP	Bonds	ρ (e/a^3)	$\nabla^2\rho$ (e/a^5)	ε	V^a
4	O2–H8	+0.0181	+0.0878	+0.5847	−0.0144
5	H8–O36	+0.0127	+0.0487	+0.0339	−0.0081
8	H5–N9	+0.0436	+0.1157	+0.0397	−0.0392
18	H16–O36	+0.0029	+0.0107	+0.0863	−0.0017
26	H23–O36	+0.0091	+0.0279	+0.0808	−0.0054
42	O2–O37	+0.0066	+0.0267	+0.2101	−0.0052
43	H19–O37	+0.0073	+0.0263	+0.1561	−0.0047
49	O35–H41	+0.0179	+0.0868	+0.6171	−0.0143
51	H38–N42	+0.0421	+0.1147	+0.0433	−0.0375
55	H41–O67	+0.0148	+0.0553	+0.1317	−0.0097
65	H49–O67	+0.0062	+0.0181	+0.0638	−0.0035
66	H33–H53	+0.0016	+0.0052	+0.1869	−0.0007
70	H56–O67	+0.0114	+0.0324	+0.0980	−0.0068

V^a (hartree.e/a^3).

HBPAH shows two different sets of HBs, intramolecular and intermolecular, with the water molecule (solvent interaction). The intramolecular HB was displayed between oxygen next to pyridine moiety and the hydrazide hydrogen, with the O-H ρ value (O2-H8 = +0.0181 e/a^3 and O35-H41 = +0.0179 e/a^3). The solvent-based HBs measure weaker than the intramolecular HB with O-H ρ values at BCPs, H49-O67, H56-O67, and H41-O67 were +0.0062 e/a^3, +0.0114 e/a^3, and +0.0148 e/a^3, respectively (Table 4 and Table S2).

3.1.3. Natural Bonding Orbital (NBO) Analysis

We next used NBO analysis to interpret charge transformation, different types of HB (inter- and intra-molecular), and hyper conjugative interactions [95–97]. For all orbitals, second-order perturbation energy $E^{(2)}$ could be calculated from Equation (9).

$$E^{(2)} = q_i \frac{\left(F_{i,j}\right)^2}{\varepsilon_j - \varepsilon_i} \quad (9)$$

qi is donor orbital occupancy, εj and εi are diagonal elements, and $F(i,j)$ is off-diagonal NBO Fock matrix element. For HBPAH, all $E^{(2)}$ values are displayed in Table S3, while the imperative $E^{(2)}$ values are arranged in Table 5.

Among probable electronic π→π* transitions of the highest magnitude, π(C46-C48)→π*(N39-C50) corresponds to stabilization energy of 30.35, *kcal/mol* in HBPAH. The transitions such as ∂(C13-H14)→∂*(C11-C13) show the lowest stabilization energy of 0.51 *kcal/mol* for HBPAH, corresponding to weak interactions between the electron donor and acceptor. Other π→π* interactions, such as π(C13-C15)→π*(N6-C17), π(N39-C50)→π*(C43-C44), π(C43-C44)→π*(C45-C48), and π(C46-C48)→π*(N39-C50), yield 29.62, 28.32, 22.45 and 30.35 *kcal/mol* stabilization energies, respectively (Table 5).

Moreover, the most prominent interactions in LP→π* manifested as LP1(N40)→π*(O36-C54), LP1(N7)→π*(O3-C21), LP2(O35)→π*(N39-C50), and LP2(Cl1)→π*(N6-C10) showed stabilization energies of 62.83, 56.48, 35.80, and 5.79 *kcal/mol*, respectively (Table 5).

For HBPAH, two additional interactions, i.e., LP1(N40)→π*(O36-C54) and LP1(N7)→π*(O3-C21) with respective high stabilization energy values of 62.83 and 56.48 kcal/mol, indicated the strong HB between lone-pair to anti-bonding orbitals in our HS and QT-AIM analyses. We conclude that these interactions directly stabilize HBPAH in its solid-state.

Table 5. Natural bonding orbital (NBO) analysis for HBPAH using the B3LYP/6-311G(d,p) level.

Compound	Donor(i)	Type	Acceptor(j)	Type	$E^a(2)$	$E(j)E(i)^b$(a.u)	$F(i,j)^c$(a.u)
HBPAH	C13-H14	∂	C11-C13	∂ *	0.51	1.09	0.021
	C46-C48	π	N39-C50	π *	30.35	0.26	0.082
	C13-C15	π	N6-C17	π *	29.62	0.26	0.082
	N39-C50	π	C43-C44	π *	28.32	0.33	0.088
	C43-C44	π	C45-C48	π *	22.45	0.30	0.074
	C30-C32	π	C26-C28	π *	21.46	0.29	0.071
	C13-C15	π	C10-C11	π *	16.09	0.27	0.060
	C43-C44	π	N39-C50	π *	14.78	0.27	0.058
	N40	LP(1)	O36-C54	π *	62.83	0.29	0.121
	N7	LP(1)	O3-C21	π *	56.48	0.29	0.117
	O35	LP(2)	N39-C50	π *	35.80	0.32	0.103
	Cl1	LP(2)	N6-C10	∂ *	5.79	0.85	0.063
	N9	LP(1)	N7-H8	∂ *	7.04	0.81	0.069
	O2	LP(1)	N6-C17	∂ *	6.61	1.08	0.076
	O36	LP(1)	N7-H8	∂ *	2.22	1.13	0.045

* antibonding energetic orbitals; $^aE^{(2)}$ is the energy of hyper conjugative interaction (stabilization energy in kcal mol^{-1}); $^bE(j)E(i)$ is the energy difference between donor and acceptor *i* and *j* NBO orbitals; $^cF(i;j)$ is the Fock matrix element between *i* and *j* NBO orbitals.

3.1.4. Natural Population Analysis (NPA)

For HBPAH, the natural population-based analysis on NBO was determined by B3LYP/6-311G(d,p) (Figure S7). The phenomenon correlates to charge transformation, and the electronegativity equalization process occurs in reaction to access the electrostatic ability on the external surfaces of the structure [66,98,99]. The charges of atoms play a crucial role within the molecular conformation and bonding capability in HBPAH. The electronegative atoms such as Cl, O, and N made unequal redistribution of the electron density over the pyridine or aromatic rings. Atomic charge of oxygen atoms was O2 (−0.38655*e*) and O36 (−0.39665*e*), and for hydrogen atoms, charges were H8(0.28089*e*) and H38(0.281296*e*), respectively, due to the involvement of these atoms in the intermolecular hydrogen bonding interactions.

Furthermore, the NPA of HBPAH showed that carbon atoms, namely C10, C17, C18, C22, C25, C43, C50, C51, C54, C55, and C58, were positively charged, while C11, C13, C15, C24, C25, C28, C30, C44, C46, C48, C57, C59, C61, C63, and C65 were negatively charged (Figure S7). Moreover, all oxygen, nitrogen, and chlorine atoms were negatively charged. All hydrogens are positively charged in HBPAH.

3.1.5. Frontier Molecular Orbital (FMO) Analysis

The FMOs evaluate chemical bond strength and molecule stability [100]. In HBPAH, the energy of HOMO, LUMO, and its two upper and lower orbitals (HOMO-1, HOMO-2, LUMO+1, LUMO+2) were calculated by the TD-DFT/B3LYP/6-311G (d, p) and displayed in Figure 10. The energy difference between HOMO-LUMO is assumed to be a significant key factor to illustrate the chemical reactivity, optical properties, kinetic stability, and electronic character of the compounds [101,102]. Table 6 shows the energy data with their energy gap (ΔE) for six MO (molecular orbitals).

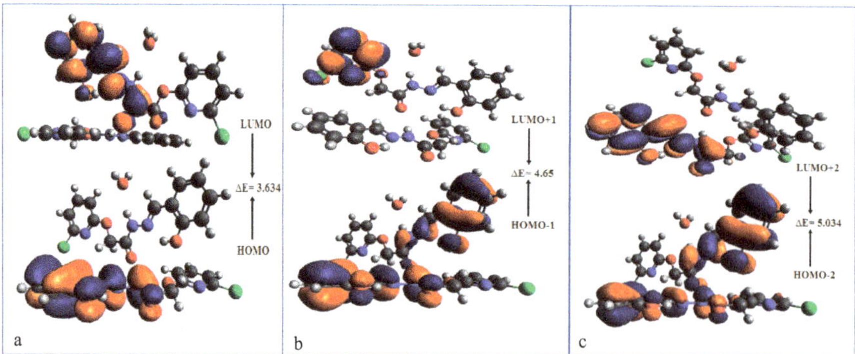

Figure 10. Frontier molecular orbitals (FMOs) for HBPAH. (**a**) E_{HOMO}, E_{LUMO}, (**b**) E_{HOMO-1}, E_{LUMO+1}, and (**c**) E_{HOMO-2}, E_{LUMO+2}, molecular orbitals. d energy gap (ΔE) in shown in *eV* of the entitled compound at the DFT/ B3LYP/6-311G (d,p) level of theory.

Table 6. The E_{HOMO}, E_{LUMO}, E_{HOMO-1}, E_{LUMO+1}, E_{HOMO-2}, E_{LUMO+2}, and energy gap (ΔE) in eV of the entitled compound at the DFT/ B3LYP/6-311G (d,p) level of theory.

HBPAH		
MO(s)	**Energy**	**ΔE(eV)**
HOMO	−5.600	3.634
LUMO	−1.966	
HOMO-1	−6.266	4.65
LUMO+1	−1.616	
HOMO-2	−6.274	5.034
LUMO+2	−1.240	

HOMO = highest occupied molecular orbital; LUMO = lowest unoccupied molecular orbital, MO = molecular orbital.

Figure 10 shows that HBPAH contained an energy gap of 3.634 eV, which exposed the effective intra-molecular charge transfer (ICT) within the compound. For HBPAH, the HOMO was populated on the first part of molecule (Z)-N′-(2-hydroxybenzylidene) acetohydrazide moiety and a small effect exists on the hydroxyl group. LUMO was populated on the second part of the molecule, i.e., (Z)-N′-(2-hydroxybenzylidene) propionohydrazide moiety (Figure 10). HOMO and LUMO energy gap values for two antifungal 1,2,4-triazolo[4,3-a]pyridine derivatives were found to be 4.318 and 3.705 *eV*, where high energy gap was associated with the more potent antifungal compound [16].

HBPAH contained an *IP* value of 5.6 eV and an *EA* value of 1.966 eV. Its electron loss and the electron gain capacity were defined by the ionization potential and the electron affinity values, which correlate to the HOMO-LUMO energy difference. Consequently, the *IP* value shows a lower magnitude than the *EA* value, indicating that HBPAH contained excellent electron-donating capability. This supports the findings of global electrophilicity (ω) (Table S4). In HBPAH, the calculated global softness (σ) values obtained were lower than the global hardness (η) values, making HBPAH stable and relatively unreactive. Additionally, the chemical potential (μ) value (−3.783 (eV)) revealed that HBPAH was chemically hard with the affective electron-donating ability and highest kinetic stability (Table S4).

3.1.6. Molecular Electrostatic Potential (MEP)

The MEP significance shows the size and configuration of the molecule, along with neutral (white), negative (red), and positive (blue) electrostatic potential regions comparable to shading assessing scheme. MEP explores the connection between molecular structural insights and physicochemical properties [103]. We Analyzed HBPAH's MEP surface through the B3LYP/6-311G(d,p) level of theory, as shown in Figure 11. The negative red indicates the electrophilic sites at the oxygen atoms. Therefore, the oxygen atoms are the most effective target for nucleophilic attack, along with the most suitable sphere to attack the molecules' positive zones. The negative potential magnitude of HBPAH is -1.00×10^{-2} to 1.00×10^{-2} *a.u.* Green areas represent the region of zero potential. The blue areas of the HBPAH molecule situate over the hydrogen atoms. They show a combination of positive charges, demonstrating the nucleophilic localities.

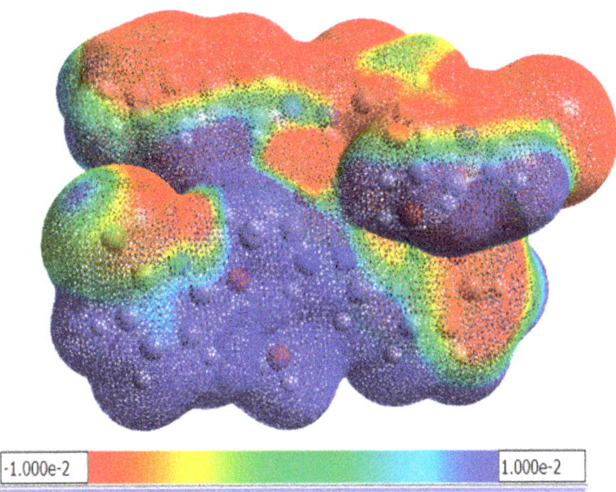

Figure 11. Molecular electrostatic potential and color scheme of HBPAH.

4. Conclusions

In conclusion, we used a room-temperature sonochemical approach to synthesize crystalline (*E*)-2-((6-chloropyridin-2-yl)oxy)-N′-(2-hydroxybenzylidene)acetohydrazide. The SC-XRD study revealed the presence of attractive intermolecular forces for the structural stabilization in this Triclinic crystal system with $P\overline{1}$ space group. The QT-AIM and Hirshfeld analysis revealed the presence of non-covalent interactions (NCIs); Scheme H5-N9, H38-N42, H16-O36, H8-O36, H19-O37, and H33-H53 that stabilize the structure of the compound. The NBO study showed that HBPAH has molecular stability of hyper-conjugation due to the intramolecular charge transfer (62.83, kcal/mol for LP1(N40) $\rightarrow\pi^*$(O36-C54)). The HOMO/LUMO energy band gap value describes the possible charge-transfer interactions, which occur inside the molecule. The calculated FMO energy bandgap of HBPAH is 3.634 eV, which illustrates it has intra-molecular charge-transferability and good NLO properties. The global reactivity descriptors calculation illustrates less reactivity and good stability. The MEP map displayed the negative red areas indicating the electrophilic sites at the oxygen atoms. All computational and experimental findings determined that HBPAH exists in stabilized crystal form because of non-covalent interactions (NCIs) and intra- and inter-molecular H-bonding interactions.

Supplementary Materials: The following are available online at http://www.mdpi.com/2073-4352/10/9/778/s1, Figure S1: 1H and 13C NMR spectra of HBPAH, Figure S2: The 1H-NMR of the title compound in (CD3)2SO-d6 showing the integration of methylenic (CH2) 1Hs, which has been used as a tool to calculate the ratio of two isomers (A and B), Figure S3: The geometrical parameters (bond lengths (Å) and bond angles (°) of entitled compound calculated through XRD and at DFT/ B3LYP/6-311G (d,p) level of theory, Figure S4: ORTEP diagram of HBPAH compound, Figure S5: Hirshfeld surfaces of the entitled compound mapped over, de and di for HBPAH (1 a.u. of electron density = 6.748 e.Å^{-3}), Figure S6: 2-D Fingerprint plots for individual contributions in HBPAH, Figure S7: Natural population analysis (NPA) of entitled compound, Table S1: Comparison of XRD and DFT values of bond length (Å) and bond angle (°) of HBPAH, Table S2: AIM properties of HBPAH; Electronic density (ρ), Laplacian of density ($\nabla^2\rho$), ellipticity (ε) and density of potential energy (V), Table S3: NBO analysis of HBPAH at DFT/ B3LYP/6-311G (d,p) level of theory, Table S4: Ionization potential (IP), electron affinity (EA), electronegativity (X), global hardness (η), chemical potential (μ), global electrophilicity index (ω) and global softness (σ).

Author Contributions: Conceptualization, A.A., C.L., and M.F.u.R.; methodology, A.A., and F.K.; software, S.A., and M.F.u.R.; validation, M.N.T., M.A.; formal analysis, C.L., A.A.; investigation, M.K., S.A.; resources, J.I., and F.K.; data curation, M.K.; writing—original draft preparation, A.A., C.L., and M.F.u.R.; writing—review and editing, F.K. and M.F.u.R.; visualization, M.N.T.; supervision, A.A., C.L.; project administration, A.A. and M.K.; funding acquisition, C.L. All authors have read and agreed to the published version of the manuscript.

Funding: Funding in the Lu lab was provided by the Shanghai Science and Technology Committee (19ZR1471100), Fundamental Research Funds for the Central Universities (19D210501, 19D310517).

Acknowledgments: The authors thank Muhammad Mustaqeem and research students in Muhammad Khalid's group: Wajeeha Anwer, Aamina Khalid, and Muhammad Zahid for their participation in terms of careful reading.

Conflicts of Interest: The authors declare no conflict of interest.

References

1. Rollas, S.; Küçükgüzel, S.G. Biological activities of hydrazone derivatives. *Molecules* **2007**, *12*, 1910–1939. [CrossRef] [PubMed]
2. Verma, G.; Marella, A.; Shaquiquzzaman, M.; Akhtar, M.; Ali, M.R.; Alam, M.M. A review exploring biological activities of hydrazones. *J. Pharm. Bioallied Sci.* **2014**, *6*, 69. [PubMed]
3. Hussain, I.; Ali, A. Exploring the Pharmacological Activities of Hydrazone Derivatives: A Review. *J. Phytochem. Biochem.* **2017**, *1*, 1–5.
4. Su, X.; Aprahamian, I. Hydrazone-based switches, metallo-assemblies and sensors. *Chem. Soc. Rev.* **2014**, *43*, 1963–1981. [CrossRef] [PubMed]
5. Rane, R.A.; Telvekar, V.N. Synthesis and evaluation of novel chloropyrrole molecules designed by molecular hybridization of common pharmacophores as potential antimicrobial agents. *Bioorg. Med. Chem. Lett.* **2010**, *20*, 5681–5685. [CrossRef] [PubMed]
6. El-Sabbagh, O.; Shabaan, M.A.; Kadry, H.H.; Al-Din, E.S. New octahydroquinazoline derivatives: Synthesis and hypotensive activity. *Eur. J. Med. Chem.* **2010**, *45*, 5390–5396. [CrossRef] [PubMed]
7. Ma, X.D.; Yang, S.Q.; Gu, S.X.; He, Q.Q.; Chen, F.E.; De Clercq, E.; Balzarini, J.; Pannecouque, C. Synthesis and anti-HIV activity of Aryl-2-[(4-cyanophenyl) amino]-4-pyrimidinone hydrazones as potent non-nucleoside reverse transcriptase inhibitors. *ChemMedChem* **2011**, *6*, 2225–2232. [CrossRef]
8. Tributino, J.L.; Duarte, C.D.; Corrêa, R.S.; Doriguetto, A.C.; Ellena, J.; Romeiro, N.C.; Castro, N.G.; Miranda, A.L.P.; Barreiro, E.J.; Fraga, C.A. Novel 6-methanesulfonamide-3, 4-methylenedioxyphenyl-N-acylhydrazones: Orally effective anti-inflammatory drug candidates. *Bioorg. Med. Chem.* **2009**, *17*, 1125–1131. [CrossRef]
9. Liu, W.-Y.; Li, H.-Y.; Zhao, B.-X.; Shin, D.-S.; Lian, S.; Miao, J.-Y. Synthesis of novel ribavirin hydrazone derivatives and anti-proliferative activity against A549 lung cancer cells. *Carbohydr. Res.* **2009**, *344*, 1270–1275. [CrossRef]
10. Gil-Longo, J.; Laguna, M.D.L.R.; Verde, I.; Castro, M.E.; Orallo, F.; Fontenla, J.A.; Calleja, J.M.; Ravina, E.; Teran, C. Pyridazine derivatives. XI: Antihypertensive activity of 3-hydrazinocycloheptyl [1, 2-c] pyridazine and its hydrazone derivatives. *J. Pharm. Sci.* **1993**, *82*, 286–290. [CrossRef]
11. Mahajan, A.; Kremer, L.; Louw, S.; Guéradel, Y.; Chibale, K.; Biot, C. Synthesis and in vitro antitubercular activity of ferrocene-based hydrazones. *Bioorg. Med. Chem. Lett.* **2011**, *21*, 2866–2868. [CrossRef] [PubMed]
12. Fattorusso, C.; Campiani, G.; Kukreja, G.; Persico, M.; Butini, S.; Romano, M.P.; Altarelli, M.; Ros, S.; Brindisi, M.; Savini, L. Design, synthesis, and structure–activity relationship studies of 4-quinolinyl-and 9-acrydinylhydrazones as potent antimalarial agents. *J. Med. Chem.* **2008**, *51*, 1333–1343. [CrossRef] [PubMed]
13. De Oliveira, K.N.; Costa, P.; Santin, J.R.; Mazzambani, L.; Bürger, C.; Mora, C.; Nunes, R.J.; De Souza, M.M. Synthesis and antidepressant-like activity evaluation of sulphonamides and sulphonyl-hydrazones. *Bioorg. Med. Chem.* **2011**, *19*, 4295–4306. [CrossRef] [PubMed]
14. Musad, E.A.; Mohamed, R.; Saeed, B.A.; Vishwanath, B.S.; Rai, K.L. Synthesis and evaluation of antioxidant and antibacterial activities of new substituted bis (1, 3, 4-oxadiazoles), 3, 5-bis (substituted) pyrazoles and isoxazoles. *Bioorg. Med. Chem. Lett.* **2011**, *21*, 3536–3540. [CrossRef]
15. Jain, J.; Kumar, Y.; Sinha, R.; Kumar, R.; Stables, J. Menthone aryl acid hydrazones: A new class of anticonvulsants. *Med. Chem.* **2011**, *7*, 56–61. [CrossRef]

16. Mu, J.-X.; Shi, Y.-X.; Wu, H.-K.; Sun, Z.-H.; Yang, M.-Y.; Liu, X.-H.; Li, B.-J. Microwave assisted synthesis, antifungal activity, DFT and SAR study of 1,2,4-triazolo[4,3-a]pyridine derivatives containing hydrazone moieties. *Chem. Cent. J.* **2016**, *10*, 50. [CrossRef]
17. Özdemir, A.; Turan-Zitouni, G.; Asim kaplancikli, Z.; Demirci, F.; Iscan, G. Studies on hydrazone derivatives as antifungal agents. *J. Enzym. Inhib. Med. Chem.* **2008**, *23*, 470–475. [CrossRef]
18. Lidström, P.; Tierney, J.; Watheyb, B.; Westmana, J. Microwave assisted organic synthesis: A review. *Tetrahedron* **2001**, *57*, 9225–9283. [CrossRef]
19. Easmon, J.; Pürstinger, G.; Thies, K.-S.; Heinisch, G.; Hofmann, J. Synthesis, structure−activity relationships, and antitumor studies of 2-benzoxazolyl hydrazones derived from alpha-(N)-acyl heteroaromatics. *J. Med. Chem.* **2006**, *49*, 6343–6350. [CrossRef]
20. Mukherjee, S.; Chowdhury, S.; Paul, A.K.; Banerjee, R. Selective extraction of palladium (II) using hydrazone ligand: A novel fluorescent sensor. *J. Lumin.* **2011**, *131*, 2342–2346. [CrossRef]
21. Chaouiki, A.; Chafiq, M.; Lgaz, H.; Al-Hadeethi, M.R.; Ali, I.H.; Masroor, S.; Chung, I.-M. Green Corrosion Inhibition of Mild Steel by Hydrazone Derivatives in 1.0 M HCl. *Coatings* **2020**, *10*, 640. [CrossRef]
22. Chaston, T.B.; Richardson, D.R. Interactions of the pyridine-2-carboxaldehyde isonicotinoyl hydrazone class of chelators with iron and DNA: Implications for toxicity in the treatment of iron overload disease. *JBIC J. Biol. Inorg. Chem.* **2003**, *8*, 427–438. [CrossRef] [PubMed]
23. Steiner, T. The whole palette of hydrogen bonds. *Angew. Chem. Int. Ed.* **2002**, *41*, 14–76.
24. Gerlt, J.A.; Kreevoy, M.M.; Cleland, W.; Frey, P.A. Understanding enzymic catalysis: The importance of short, strong hydrogen bonds. *Chem. Biol.* **1997**, *4*, 259–267. [CrossRef]
25. Perrin, C.L.; Nielson, J.B. "Strong" hydrogen bonds in chemistry and biology. *Annu. Rev. Phys. Chem.* **1997**, *48*, 511–544. [CrossRef]
26. Emamian, S.; Lu, T.; Kruse, H.; Emamian, H. Exploring Nature and Predicting Strength of Hydrogen Bonds: A Correlation Analysis Between Atoms-in-Molecules Descriptors, Binding Energies, and Energy Components of Symmetry-Adapted Perturbation Theory. *J. Comput. Chem.* **2019**, *40*, 2868–2881. [CrossRef]
27. Zundel, G. Hydrogen bonds with large proton polarizability and proton transfer processes in electrochemistry and biology. *Adv. Chem. Phys.* **1999**, *111*, 1–217. [CrossRef]
28. Bernstein, J.; Etter, M.C.; Leiserowitz, L. The role of hydrogen bonding in molecular assemblies. *Struct. Correl.* **1994**, 431–507. [CrossRef]
29. Pokorna, P.; Krepl, M.; Kruse, H.; Sponer, J. MD and QM/MM Study of the Quaternary HutP Homohexamer Complex with mRNA, L-Histidine Ligand, and Mg^{2+}. *J. Chem. Theory Comput.* **2017**, *13*, 5658–5670. [CrossRef]
30. Zhang, Z.; Schreiner, P.R. (Thio) urea organocatalysis—What can be learnt from anion recognition? *Chem. Soc. Rev.* **2009**, *38*, 1187–1198. [CrossRef]
31. Doyle, A.G.; Jacobsen, E.N. Small-molecule H-bond donors in asymmetric catalysis. *Chem. Rev.* **2007**, *107*, 5713–5743. [CrossRef] [PubMed]
32. Taylor, M.S.; Jacobsen, E.N. Asymmetric catalysis by chiral hydrogen-bond donors. *Angew. Chem. Int. Ed.* **2006**, *45*, 1520–1543. [CrossRef] [PubMed]
33. Schreiner, P.R. Metal-free organocatalysis through explicit hydrogen bonding interactions. *Chem. Soc. Rev.* **2003**, *32*, 289–296. [CrossRef]
34. Pihko, P.M. Activation of carbonyl compounds by double hydrogen bonding: An emerging tool in asymmetric catalysis. *Angew. Chem. Int. Ed.* **2004**, *43*, 2062–2064. [CrossRef] [PubMed]
35. Ali, A.; Khalid, M.; Abid, S.; Iqbal, J.; Tahir, M.N.; Rauf Raza, A.; Zukerman-Schpector, J.; Paixão, M.W. Facile synthesis, crystal growth, characterization and computational study of new pyridine-based halogenated hydrazones: Unveiling the stabilization behavior in terms of noncovalent interactions. *Appl. Organomet. Chem.* **2020**, *34*, e5399. [CrossRef]
36. Zhang, L.; Yang, M.; Hu, B.; Sun, Z.; Liu, X.; Weng, J.; Tan, C. Microwave-assisted synthesis of novel 8-chloro-[1, 2, 4] triazolo [4, 3- alpha] pyridine derivatives. *Turk. J. Chem.* **2015**, *39*, 867–873. [CrossRef]
37. De la Hoz, A.; Diaz-Ortiz, A.; Moreno, A. Microwaves in organic synthesis. Thermal and non-thermal microwave effects. *Chem. Soc. Rev.* **2005**, *34*, 164–178. [CrossRef] [PubMed]
38. Amariucai-Mantu, D.; Mangalagiu, V.; Danac, R.; Mangalagiu, I.I. Microwave Assisted Reactions of Azaheterocycles Formedicinal Chemistry Applications. *Molecules* **2020**, *25*, 716. [CrossRef]
39. *APEX2 (Version 1.22) and SAINT-Plus (Version 7.06a)*; Bruker: Billica, MA, USA, 2009.

40. Higashino, T.; Akiyama, Y.; Kojima, H.; Kawamoto, T.; Mori, T. Organic semiconductors and conductors with tert-butyl substituents. *Crystals* **2012**, *2*, 1222–1238. [CrossRef]
41. Sheldrick, G.M. Crystal structure refinement with SHELXL. *Acta Crystallogr. C Struct. Chem.* **2015**, *71*, 3–8. [CrossRef]
42. Delgado, G.; Henao, J.; Quintana, J.; Al-Maqtari, H.; Jamalis, J.; Sirat, H. Structural Characterization of a New Chalcone Compound Containing a Thiophene Moiety:(E)-3-(5-Bromothiophen-2-YL)-1-(2, 5-Dichlorothiophen-3-YL)-2-Propen-1-One. *J. Struct. Chem.* **2018**, *59*, 1440–1445. [CrossRef]
43. Thomassen, I.K.; McCormick, L.J.; Ghosh, A. Molecular Structure of a Free-Base β-Octaiodo-meso-tetraarylporphyrin. A Rational Route to cis Porphyrin Tautomers? *Cryst. Growth Des.* **2018**, *18*, 4257–4259. [CrossRef]
44. Ali, A.; Badawy, M.; Shah, R.; Rehman, W.; El kilany, Y.; El Ashry, E.S.H.; Tahir, N. Synthesis, characterization and in-silico admet screening of mono-and dicarbomethoxylated 6, 6′-methylenebis (2-cyclohexyl-4-methylphenol) and their hydrazides and hydrazones. *Chim. Sin.* **2017**, *8*, 446–460.
45. Braga, A.A.C.; Ujaque, G.; Maseras, F. A DFT Study of the Full Catalytic Cycle of the Suzuki–Miyaura Cross-Coupling on a Model System. *Organometallics* **2006**, *25*, 3647–3658. [CrossRef]
46. García-Melchor, M.; Braga, A.A.C.; Lledós, A.; Ujaque, G.; Maseras, F. Computational Perspective on Pd-Catalyzed C–C Cross-Coupling Reaction Mechanisms. *Acc. Chem. Res.* **2013**, *46*, 2626–2634. [CrossRef]
47. Braga, A.A.; Morgon, N.H.; Ujaque, G.; Maseras, F. Computational characterization of the role of the base in the Suzuki–Miyaura cross-coupling reaction. *J. Am. Chem. Soc.* **2005**, *127*, 9298–9307. [CrossRef]
48. Frisch, M.J.; Trucks, G.W.; Schlegel, H.B.; Scuseria, G.; Robb, M.A.; Cheeseman, J.R.; Scalmani, G.; Barone, V.; Mennucci, B.; Petersson, G.; et al. *D. 0109, Revision D. 01*; Gaussian Inc.: Wallingford, CT, USA, 2009.
49. Dennington, R.; Keith, T.; Millam, J. *GaussView*; Version 5; Semichem Inc.: Shawnee Mission, KS, USA, 2009.
50. Chemcraft. Graphical Software for Visualization of Quantum Chemistry Computations. Available online: https://www.chemcraftprog.com (accessed on 31 March 2020).
51. Hanwell, M.D.; Curtis, D.E.; Lonie, D.C.; Vandermeersch, T.; Zurek, E.; Hutchison, G.R. Avogadro: An advanced semantic chemical editor, visualization, and analysis platform. *J. Cheminform.* **2012**, *4*, 17. [CrossRef]
52. O'boyle, N.M.; Tenderholt, A.L.; Langner, K.M. Cclib: A library for package-independent computational chemistry algorithms. *J. Comput. Chem.* **2008**, *29*, 839–845. [CrossRef]
53. Keith, T.A. *AIMAll, TK Gristmill Software*; AIMAll: Overland Park, KS, USA, 2012.
54. Wolff, S.; Grimwood, D.; McKinnon, J.; Turner, M.; Jayatilaka, D.; Spackman, M. *CrystalExplorer*; Version 3.1; University of Western Australia: Crawley, Australia, 2012.
55. Glendenning, E.; Reed, A.; Carpenter, J.; Weinhold, F. *NBO*; Version 3.1; 2001. Available online: http://www.ccl.net/cca/software/MS-WIN95-NT/mopac6/nbo.htm (accessed on 31 August 2020).
56. Weinhold, F.; Glendening, E.D. *NBO 5.0 Program Manual*; Theoretical Chemistry Institute and Department of Chemistry, University of Wisconsin: Madison, WI, USA, 2001; Volume 53706, p. 101.
57. Gross, E.K.U.; Kohn, W. Time-dependent density-functional theory. In *Advances in Quantum Chemistry*; Löwdin, P.-O., Ed.; Academic Press: Cambridge, MA, USA, 1990; Volume 21, pp. 255–291.
58. Burke, K.; Werschnik, J.; Gross, E. Time-dependent density functional theory: Past, present, and future. *J. Chem. Phys.* **2005**, *123*, 062206. [CrossRef]
59. Spackman, M.A.; Jayatilaka, D. Hirshfeld surface analysis. *CrystEngComm* **2009**, *11*, 19–32. [CrossRef]
60. McKinnon, J.J.; Jayatilaka, D.; Spackman, M.A. Towards quantitative analysis of intermolecular interactions with Hirshfeld surfaces. *Chem. Commun.* **2007**, 3814–3816. [CrossRef] [PubMed]
61. Kumar, P.S.V.; Raghavendra, V.; Subramanian, V. Bader's theory of atoms in molecules (AIM) and its applications to chemical bonding. *J. Chem. Sci.* **2016**, *128*, 1527–1536. [CrossRef]
62. Parr, R.G.; Donnelly, R.A.; Levy, M.; Palke, W.E. Electronegativity: The density functional viewpoint. *J. Chem. Phys.* **1978**, *68*, 3801–3807. [CrossRef]
63. Chattaraj, P.K.; Sarkar, U.; Roy, D.R. Electrophilicity Index. *Chem. Rev.* **2006**, *106*, 2065–2091. [CrossRef] [PubMed]
64. Chattaraj, P.K.; Roy, D.R. Update 1 of: Electrophilicity Index. *Chem. Rev.* **2007**, *107*, PR46–PR74. [CrossRef]
65. Saravanan, S.; Balachandran, V. Quantum chemical studies, natural bond orbital analysis and thermodynamic function of 2, 5-dichlorophenylisocyanate. *Spectrochim. Acta Part A Mol. Biomol. Spectrosc.* **2014**, *120*, 351–364. [CrossRef]

66. Parr, R.G.; Pearson, R.G. Absolute hardness: Companion parameter to absolute electronegativity. *J. Am. Chem. Soc.* **1983**, *105*, 7512–7516. [CrossRef]
67. Sheela, N.; Muthu, S.; Sampathkrishnan, S. Molecular orbital studies (hardness, chemical potential and electrophilicity), vibrational investigation and theoretical NBO analysis of 4-4′-(1H-1, 2, 4-triazol-1-yl methylene) dibenzonitrile based on abinitio and DFT methods. *Spectrochim. Acta Part A Mol. Biomol. Spectrosc.* **2014**, *120*, 237–251. [CrossRef]
68. Parthasarathi, R.; Padmanabhan, J.; Elango, M.; Subramanian, V.; Chattaraj, P. Intermolecular reactivity through the generalized philicity concept. *Chem. Phys. Lett.* **2004**, *394*, 225–230. [CrossRef]
69. Politzer, P.; Truhlar, D.G. Introduction: The role of the electrostatic potential in chemistry. In *Chemical Applications of Atomic and Molecular Electrostatic Potentials*; Springer: Boston, MA, USA, 1981; pp. 1–6. [CrossRef]
70. Parr, R.G.; Szentpaly, L.v.; Liu, S. Electrophilicity index. *J. Am. Chem. Soc.* **1999**, *121*, 1922–1924. [CrossRef]
71. Roy, D.; Sarkar, U.; Chattaraj, P.; Mitra, A.; Padmanabhan, J.; Parthasarathi, R.; Subramanian, V.; Van Damme, S.; Bultinck, P. Analyzing toxicity through electrophilicity. *Mol. Divers.* **2006**, *10*, 119–131. [CrossRef] [PubMed]
72. Lesar, A.; Milošev, I. Density functional study of the corrosion inhibition properties of 1, 2, 4-triazole and its amino derivatives. *Chem. Phys. Lett.* **2009**, *483*, 198–203. [CrossRef]
73. Koopmans, T. Über die Zuordnung von Wellenfunktionen und Eigenwerten zu den einzelnen Elektronen eines Atoms. *Physica* **1934**, *1*, 104–113. [CrossRef]
74. Desiraju, G.R. The CH. cntdot. cntdot. cntdot. O hydrogen bond in crystals: What is it? *Acc. Chem. Res.* **1991**, *24*, 290–296. [CrossRef]
75. Desiraju, G.R.; Kishan, K.R. Crystal chemistry of some (alkoxyphenyl) propiolic acids. The role of oxygen and hydrogen atoms in determining stack structures of planar aromatic compounds. *J. Am. Chem. Soc.* **1989**, *111*, 4838–4843. [CrossRef]
76. Çakmak, O.; Ökten, S.; Alımlı, D.; Ersanlı, C.C.; Taslimi, P.; Koçyiğit, Ü.M. Novel piperazine and morpholine substituted quinolines: Selective synthesis through activation of 3, 6, 8-tribromoquinoline, characterization and their some metabolic enzymes inhibition potentials. *J. Mol. Struct.* **2020**, 128666. [CrossRef]
77. Spackman, M.A.; Byrom, P.G. A novel definition of a molecule in a crystal. *Chem. Phys. Lett.* **1997**, *267*, 215–220. [CrossRef]
78. Zukerman-Schpector, J.; Dallasta Pedroso, S.; Sousa Madureira, L.; Weber Paixão, M.; Ali, A.; Tiekink, E.R. 4-Benzyl-1-(4-nitrophenyl)-1H-1, 2, 3-triazole: Crystal structure and Hirshfeld analysis. *Acta Crystallogr. Sect. E Crystallogr. Commun.* **2017**, *73*, 1716–1720. [CrossRef]
79. McKinnon, J.J.; Spackman, M.A.; Mitchell, A.S. Novel tools for visualizing and exploring intermolecular interactions in molecular crystals. *Acta Crystallogr. Sect. B Struct. Sci.* **2004**, *60*, 627–668. [CrossRef]
80. Wang, H.; Yin, Z. Crystal structure and Hirshfeld surface analysis of dibutyl 5, 5′-(pentane-3, 3-diyl) bis (1H-pyrrole-5-carboxylate). *Acta Crystallogr. Sect. E Crystallogr. Commun.* **2019**, *75*, 711. [CrossRef]
81. Hirshfeld, F.L. Bonded-atom fragments for describing molecular charge densities. *Theor. Chim. Acta* **1977**, *44*, 129–138. [CrossRef]
82. Pook, N.-P.; Adam, A.; Gjikaj, M. Crystal structure and Hirshfeld surface analysis of (μ-2-{4-[(carboxylatomethyl) carbamoyl] benzamido} acetato-κ2O: O′) bis [bis (1, 10-phenanthroline-κ2N, N′) copper (II)] dinitrate N, N′-(1, 4-phenylenedicarbonyl) diglycine monosolvate octahydrate. *Acta Crystallogr. Sect. E Crystallogr. Commun.* **2019**, *75*, 667–674. [CrossRef] [PubMed]
83. Sangeetha, K.; Rajina, S.; Marchewka, M.; Binoy, J. The study of inter and intramolecular hydrogen bonds of NLO crystal melaminium hydrogen malonate using DFT simulation, AIM analysis and Hirshfeld surface analysis. *Mater. Today Proc.* **2020**, *25*, 307–315. [CrossRef]
84. Toronto, S.T. *Artificial Intelligence for Engineering Design, Analysis and Manufacturing*; AIE, Cambridge University Press: Cambridge, UK, 1989; Volume 3, pp. B1–B8. [CrossRef]
85. Koenderink, J.J.; Van Doorn, A.J. The singularities of the visual mapping. *Biol. Cybern.* **1976**, *24*, 51–59. [CrossRef] [PubMed]
86. Ali, A.; Zukerman-Schpector, J.; Weber Paixão, M.; Jotani, M.M.; Tiekink, E.R. 7-Methyl-5-[(4-methylbenzene) sulfonyl]-2H, 5H-[1, 3] dioxolo [4, 5-f] indole: Crystal structure and Hirshfeld analysis. *Acta Crystallogr. Sect. E Crystallogr. Commun.* **2018**, *74*, 184–188. [CrossRef] [PubMed]

87. Sebbar, N.K.; Hni, B.; Hökelek, T.; Jaouhar, A.; Labd Taha, M.; Mague, J.T.; Essassi, E.M. Crystal structure, Hirshfeld surface analysis and interaction energy and DFT studies of 3-{(2Z)-2-[(2, 4-dichlorophenyl) methylidene]-3-oxo-3, 4-dihydro-2H-1, 4-benzothiazin-4-yl} propanenitrile. *Acta Crystallogr. Sect. E Crystallogr. Commun.* **2019**, *75*, 721–727. [CrossRef]
88. Tahir, M.N.; Ashfaq, M.; Alexander, F.; Caballero, J.; Hernández-Rodríguez, E.W.; Ali, A. Rationalizing the stability and interactions of 2, 4-diamino-5-(4-chlorophenyl)-6-ethylpyrimidin-1-ium 2-hydroxy-3, 5-dinitrobenzoate salt. *J. Mol. Struct.* **2019**, *1193*, 185–194. [CrossRef]
89. Khalid, M.; Ali, A.; Rehman, M.F.U.; Mustaqeem, M.; Ali, S.; Khan, M.U.; Asim, S.; Ahmad, N.; Saleem, M. Exploration of Noncovalent Interactions, Chemical Reactivity, and Nonlinear Optical Properties of Piperidone Derivatives: A Concise Theoretical Approach. *ACS Omega* **2020**, *5*, 13236–13249. [CrossRef]
90. Thakur, T.S.; Dubey, R.; Desiraju, G.R. Intermolecular atom–atom bonds in crystals–a chemical perspective. *IUCrJ* **2015**, *2*, 159–160. [CrossRef]
91. Bader, R.F. Atoms in molecules. *Acc. Chem. Res.* **1985**, *18*, 9–15. [CrossRef]
92. Bader, R.; Nguyen-Dang, T. Quantum theory of atoms in molecules–Dalton revisited. In *Advances in Quantum Chemistry*; Academic Press: Cambridge, MA, USA; Elsevier: Amsterdam, The Netherlands, 1981; Volume 14, pp. 63–124. [CrossRef]
93. Desiraju, G.R. Crystal engineering: A brief overview. *J. Chem. Sci.* **2010**, *122*, 667–675. [CrossRef]
94. Hernández-Paredes, J.; Carrillo-Torres, R.C.; López-Zavala, A.A.; Sotelo-Mundo, R.R.; Hernández-Negrete, O.; Ramírez, J.Z.; Alvarez-Ramos, M.E. Molecular structure, hydrogen-bonding patterns and topological analysis (QTAIM and NCI) of 5-methoxy-2-nitroaniline and 5-methoxy-2-nitroaniline with 2-amino-5-nitropyridine (1: 1) co-crystal. *J. Mol. Struct.* **2016**, *1119*, 505–516. [CrossRef]
95. Tamer, Ö.; Avcı, D.; Atalay, Y. Quantum chemical characterization of N-(2-hydroxybenzylidene) acetohydrazide (HBAH): A detailed vibrational and NLO analysis. *Spectrochim. Acta Part A Mol. Biomol. Spectrosc.* **2014**, *117*, 78–86. [CrossRef]
96. Szafran, M.; Komasa, A.; Bartoszak-Adamska, E. Crystal and molecular structure of 4-carboxypiperidinium chloride (4-piperidinecarboxylic acid hydrochloride). *J. Mol. Struct.* **2007**, *827*, 101–107. [CrossRef]
97. James, C.; Raj, A.A.; Reghunathan, R.; Jayakumar, V.; Joe, I.H. Structural conformation and vibrational spectroscopic studies of 2, 6-bis (p-N, N-dimethyl benzylidene) cyclohexanone using density functional theory. *J. Raman Spectrosc. Int. J. Orig. Work All Asp. Raman Spectrosc. Incl. High. Order Process. Also Brillouin Rayleigh Scatt.* **2006**, *37*, 1381–1392. [CrossRef]
98. Pauling, L. *The Nature of the Chemical Bond*; Cornell University Press: Ithaca, NY, USA, 1960; Volume 260.
99. Mulliken, R. Overlap populations, bond orders and covalent bond energies. *J. Chem. Phys.* **1955**, *23*, 1841–1846. [CrossRef]
100. Javed, I.; Khurshid, A.; Arshad, M.N.; Wang, Y. Photophysical and electrochemical properties and temperature dependent geometrical isomerism in alkyl quinacridonediimines. *New J. Chem.* **2014**, *38*, 752–761. [CrossRef]
101. Khalid, M.; Ali, A.; Jawaria, R.; Asghar, M.A.; Asim, S.; Khan, M.U.; Hussain, R.; ur Rehman, M.F.; Ennis, C.J.; Akram, M.S. First principles study of electronic and nonlinear optical properties of A–D–π–A and D–A–D–π–A configured compounds containing novel quinoline–carbazole derivatives. *RSC Adv.* **2020**, *10*, 22273–22283. [CrossRef]
102. Ali, A.; Khalid, M.; Rehman, M.F.u.; Haq, S.; Ali, A.; Tahir, M.N.; Ashfaq, M.; Rasool, F.; Braga, A.A.C. Efficient Synthesis, SC-XRD, and Theoretical Studies of O-Benzenesulfonylated Pyrimidines: Role of Noncovalent Interaction Influence in Their Supramolecular Network. *ACS Omega* **2020**, *5*, 15115–15128. [CrossRef]
103. Luque, F.J.; López, J.M.; Orozco, M. Perspective on "Electrostatic interactions of a solute with a continuum. A direct utilization of ab initio molecular potentials for the prevision of solvent effects". *Theor. Chem. Acc.* **2000**, *103*, 343–345. [CrossRef]

Opinion

IsoStar Program Suite for Studies of Noncovalent Interactions in Crystals of Chemical Compounds

Alexander S. Novikov [1,2]

[1] Institute of Chemistry, Saint Petersburg State University, Universitetskaya Nab., 7/9, 199034 Saint Petersburg, Russia; a.s.novikov@spbu.ru

[2] Infochemistry Scientific Center, ITMO University, Kronverksky Pr., 49, Bldg. A, 197101 Saint Petersburg, Russia

Abstract: Both Cambridge Crystallographic Data Centre (CCDC) and Protein Data Bank (PDB) provide unique opportunities for finding information about the structures of chemical and biochemical compounds in the solid state. The IsoStar—a knowledge-based library of intermolecular interactions—is a very powerful tool for automatic analysis of a large amount of data from these databases. The IsoStar program suite could help chemists in understanding of probability of occurrence (frequencies) and spatial characteristics (directionalities) of noncovalent contacts (including hydrogen, halogen, and chalcogen bonds, as well as interactions involving π-systems) between pairs of chemical functional groups; this web application may also be useful for crystal engineers, crystallographers, medicinal chemists, and researchers in fields of computational chemistry and molecular modeling.

Keywords: IsoStar; Cambridge Structural Database; Protein Data Bank; noncovalent interactions

Citation: Novikov, A.S. IsoStar Program Suite for Studies of Noncovalent Interactions in Crystals of Chemical Compounds. *Crystals* **2021**, *11*, 162. https://doi.org/10.3390/cryst11020162

Academic Editor: Ana M. Garcia-Deibe

Received: 30 January 2021
Accepted: 5 February 2021
Published: 6 February 2021

Obviously, the most accurate and objective idea about the structure of chemical compounds in the solid state can be obtained using X-ray crystallography. To date, there are two most extensive online databases on the structure of chemical compounds: Cambridge Crystallographic Data Centre (CCDC, https://www.ccdc.cam.ac.uk) and Protein Data Bank (PDB, http://www.wwpdb.org). The CCDC is a crystallographic organization focused on small molecule crystal structures (mainly organic and organometallic compounds), whereas PDB is a database for the three-dimensional structural data of large biological molecules, such as proteins and nucleic acids. Both these databases contain almost unlimited information about the different kinds of noncovalent contacts (including hydrogen, halogen, and chalcogen bonds, as well as interactions involving π-systems) between atoms and functional groups, and this information could be potentially ultimately interesting for crystal engineers, crystallographers, medicinal chemists, and researchers in fields of computational chemistry and molecular modelling. Unfortunately, manual extraction of the information about the noncovalent contacts from these databases could be very time consuming and subjective.

However, currently there is a brilliant automatic web application for these purposes—IsoStar program suite (https://isostar.ccdc.cam.ac.uk/html/isostar.html) [1,2] (Figure 1). Crystallographic information in IsoStar is presented in the form of three-dimensional scatterplots, which could be future converted to contoured density surfaces. Each scatterplot constructed based on the CCDC or PDB search for weak (long) contacts between a pair of functional groups X and Y and demonstrates the experimentally observed distribution of X (contact group) around Y (central group). Such scatterplots provide ideas about the probability of occurrence (frequencies) and spatial characteristics (directionalities) of noncovalent contacts between pairs of chemical functional groups (e.g., C, N, O, S, Si, P, Se, or H atom, hydrophobic groups like Me or Ph, halogen atoms, and amino acids). The IsoStar program suite also provide huge statistical information (e.g., the number of crystal structures that have both the central and contact groups present, the number of crystal structures in which the central group and contact group form a contact with distance less than the sum of van der Waals radii).

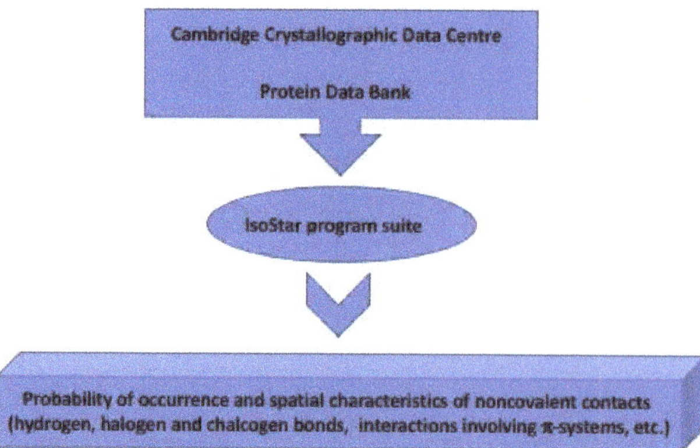

Figure 1. IsoStar program suite—an automatic web application for studies of noncovalent interactions in crystals of chemical compounds.

Inspection of the literature data reveals that the IsoStar program suite could be successfully used for studies of extended supramolecular architectures of co-crystals (up to four- and five-component assemblies) [3], smart and predictable design of molecular crystals via tunable site-specific intermolecular interactions, which provide hierarchical self-assembly [4–6], screening of metal–organic frameworks for materials discovery [7], and density functional theory supported analysis to assess the utility of σ- and π-hole interactions for crystal engineering [8].

Thus, I call the scientific community to pay attention to this useful tool for studying noncovalent interactions in the solid phase and widely use it in daily research work.

Funding: This research received no external funding.

Conflicts of Interest: The author declares no conflict of interest.

References

1. Bruno, I.J.; Cole, J.C.; Lommerse, J.P.M.; Rowland, R.S.; Taylor, R.; Verdonk, M.L. IsoStar: A Library of Information About Nonbonded Interactions. *J. Comput. Aided Mol. Des.* **1997**, *11*, 525–537. [CrossRef] [PubMed]
2. Battle, G.M.; Allen, F.H. Learning about Intermolecular Interactions from the Cambridge Structural Database. *J. Chem. Educ.* **2012**, *89*, 38–44. [CrossRef]
3. Aakeröy, C.B.; Sinha, A.S. Chapter 1: Co-crystals: Introduction and Scope. In *Co-crystals: Preparation, Characterization and Applications*; Aakeröy, C.B., Sinha, A.S., Eds.; The Royal Society of Chemistry: London, UK, 2018; pp. 1–32. [CrossRef]
4. Sinha, A.S.; Aakeröy, C.B. Design of Molecular Crystals Supramolecular Synthons. In *Comprehensive Supramolecular Chemistry II*; Elsevier: Amsterdam, The Netherlands, 2017; pp. 3–24. [CrossRef]
5. Corpinot, M.K.; Bučar, D.-K. A Practical Guide to the Design of Molecular Crystals. *Cryst. Growth Des.* **2019**, *19*, 1426–1453. [CrossRef]
6. Vologzhanina, A.V. Intermolecular Interactions in Functional Crystalline Materials: From Data to Knowledge. *Crystals* **2019**, *9*, 478. [CrossRef]
7. Moghadam, P.Z.; Li, A.; Wiggin, S.B.; Tao, A.; Maloney, A.G.P.; Wood, P.A.; Ward, S.C.; Fairen-Jimenez, D. Development of a Cambridge Structural Database Subset: A Collection of Metal–Organic Frameworks for Past, Present, and Future. *Chem. Mater.* **2017**, *29*, 2618–2625. [CrossRef]
8. Mooibroek, T.J. DFT and IsoStar Analyses to Assess the Utility of σ- and π-Hole Interactions for Crystal Engineering. *ChemPhysChem* **2021**, *22*, 141–153. [CrossRef] [PubMed]

MDPI
St. Alban-Anlage 66
4052 Basel
Switzerland
Tel. +41 61 683 77 34
Fax +41 61 302 89 18
www.mdpi.com

Crystals Editorial Office
E-mail: crystals@mdpi.com
www.mdpi.com/journal/crystals

www.ingramcontent.com/pod-product-compliance
Lightning Source LLC
LaVergne TN
LVHW070609170726
843515LV00004B/28

* 9 7 8 3 0 3 6 5 3 2 8 8 2 *